Tunisia in a Changing Climate

Assessment and Actions for Increased Resilience and Development

Photograph by Dorte Verner

Edited by Dorte Verner

THE WORLD BANK
Washington, D.C.

ISBN (paper): 978-0-8213-9857-9
ISBN (electronic): 978-0-8213-9858-6
DOI: 10.1596/978-0-8213-9857-9

Cover photo credit: © Dorte Verner, used with permission.

Library of Congress Cataloging-in-Publication Data.

Verner, Dorte.
  Tunisia in a changing climate : assessment and actions for increased resilience and development/
  Dorte Verner; Sustainable Development Department, Middle East and North Africa Region.
    pages cm
  Includes bibliographical references.
  ISBN 978-0-8213-9857-9 — ISBN 978-0-8213-9858-6 (ebook)
  1. Climatic changes—Tunisia.   2. Climatic changes—Economic aspects—Tunisia.
  3. Climatic changes—Social aspects—Tunisia. I. World Bank. Middle East and
North Africa Region. Sustainable Development. II. Title.
  QC991.T8V47 2013
  363.738'7409611—dc23
                                                                              2012049875

Contents

Figures

Tables

Map of Tunisia

IBRD 33500R1

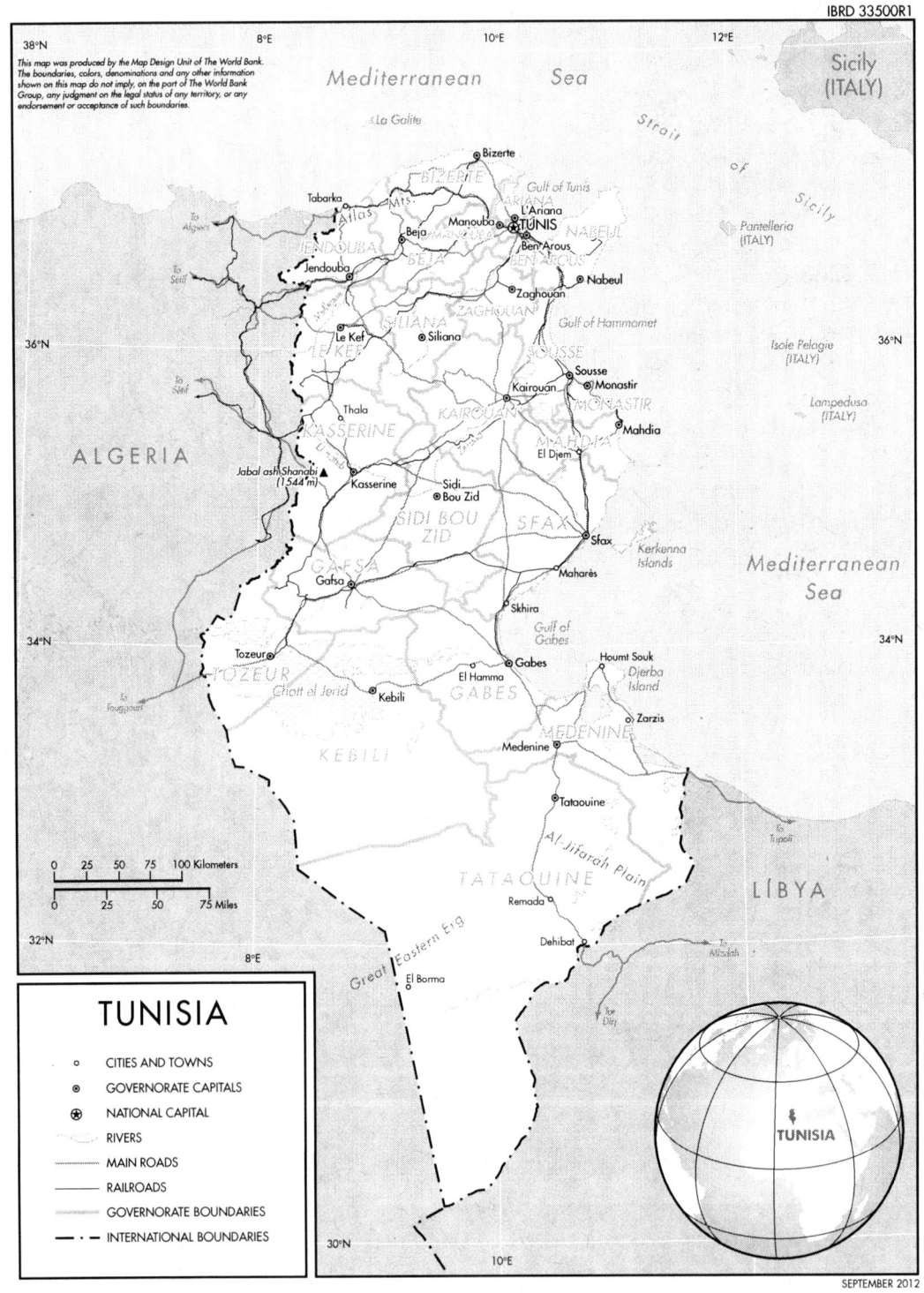

SEPTEMBER 2012

Preface

The Government of Tunisia (GoT) has requested the assistance of the World Bank in providing analytical and technical inputs in order to address critical challenges facing development in a changing climate. Based on this request, the World Bank produced this report *Tunisia in a Changing Climate* in partnership with the Ministry of Environment with the objective of providing information on climate change and development in Tunisia, including technical guidance on adaptation to climate change.

Tunisia is and will continue be impacted by climate variability and change mainly through the adverse effects resulting from increasing temperatures, reduced and variable precipitation, and sea level rise through, for example, salt water intrusion. Climate change impacts are projected to increase water scarcity, the frequency of droughts and flooding. These impacts in turn negatively affect livelihoods and human well-being especially of vulnerable people and sectors, and Tunisia's arid and coastal areas in particular. The projected negative impacts of climate change call for both adaptation measures to increase resilience and mitigation measures to curb greenhouse gas emissions.

In this context, the GoT is working to develop national studies, strategies, action plans, and project portfolios for key sectors with respect to climate change mitigation and adaptation. The World Bank plays a complementing role by providing technical assistance through this task and by supporting investments in the agriculture, health, energy, water, and urban sectors, among others. This report is responsive to the objectives of the World Bank FY13-FY14 Interim Strategy for Tunisia, in its three areas of engagement: (1) laying the foundation for sustainable growth and job creation; (2) promoting social and economic inclusion; and (3) strengthening governance: voice, transparency, and accountability. *Tunisia in a Changing Climate* complements the sector assessments and action plans completed by the government by adding economic and regionally holistic analyses.

The policy options to address climate variability and change and its impacts on Tunisia were developed together with the GoT, led by the Ministry of Environment, and with Planning, Meteorological, and Regional organizations and institutes, and other stakeholder in a participatory manner.

Acknowledgments

The report was developed and managed by Dorte Verner. The team is grateful to the authors of the chapters: Chapter 1: Dorte Verner; Chapter 2: Robert Wilby; Chapter 3: Clemens Breisinger, Perrihan Al-Riffai, Richard Robertson, and Manfred Wiebelt; Chapter 4: Jakob Kronik and Viviane Clément; and Chapter 5: Dorte Verner, Tamara Levine, Viviane Clément, Rob Wilby, Clemens Breisinger, Jakob Kronik, Ferhat Esen, and Philippe Roos.

The team is grateful for the ideas and contributions provided by the Ministry of Environment of Tunisia including H.E. Mémia Benna Zayani, Minister of Environment; Habib Ben Moussa (Director General for Environment and Quality of Life (DGEQV) in the Ministry of Environment); Imed Fadhel (National Focal Point for the UNFCCC in the Ministry of Environment); Nabil Hamdi (General Direction for Sustainable Development in the Ministry of Environment); Amel Akremi (DGEQV); Awatef Messai (DGEQV); Hédi Shili (DGEQV); Ridha Guesmi (Regional Representation of the DGEQV in Tozeur); Youssef Mansouri (Regional Representation of the DGEQV in Kairouan); Kamel Aloui (General Direction for Forests in the Ministry of Agriculture); Sahla Mizgani (General Direction for Agricultural Production in the Ministry of Agriculture); Faliez Msallem (Regional Commissariat for Agricultural Development [CRDA] of Tataouine); Abd Elmajid Bouchahoua (Tataouine Governorate); Ezzedine Nasri (CRDA of Kasserine); Mustapha Taghouti (CRDA of Kasserine); Ezzedine Dalhoumi (Kasserine Gorvernorate); Zied Askri (CRDA of Tozeur); Riadh Ribh (CRDA of Kébili); Abd Elmajid Abess (Kébili Governorate); Fethi Akroute (Gabès Governorate); Ahmed Rathouan Rdhounia (CRDA of Gafsa); Ines Dakhili (General Direction for Hygiene and Environmental Protection in the Ministry of Public Health); Halima Traya (Ministry of Industry); Moncef Miled (Ministry of Regional Development and Planning); Sinan Bacha (National Center for Cartography and Teledetection); Lamine Aouni (National Center for Cartography and Teledetection); Mouna Besbes (National Agency for Energy); Yadh Labbene (National Meteorological Institute); Soumaya Ben Rached (National Meteorological Institute); and Hadhemi Kasraoui (Agency for Coastal Protection and Planning).

The team is grateful for peer-review comments from Ana Elisa Bucher, Christophe Crepin, and Svetlana Edmeades. Additionally the team has benefited from guidance from Simon Gray, Junaid Kamal Ahmad, Mats Karlsson, Laszlo Lovei, Eileen Murray, Eavan O'Halloran, Luis Constantino, and Hoonae Kim.

The core World Bank team working on the task included Viviane Clement, Tamara Levine, Dylan Murray, and Dorte Verner. The team also benefited from support from Marie Francoise How Yew Kin and Indra Raja. The team is grateful for editorial assistance by Hilary Gopnik and photo-editorial guidance and advice from Eliot Cohen. The team is grateful for the assistance from Inger Brisson and Abdia Mohamed for efficiently managing the publication process. The team is grateful for the funding received from TFESSD.

Photograph by Dorte Verner

Executive Summary

The revolution of January 2011 has created significant change in Tunisia resulting in new challenges and opportunities for addressing climate change. On the one hand, open access to information, more accountable institutions, a commission to investigate corruption, new laws, and a new constitution may enhance governance and transparency essential for effective climate change responses. On the other hand, political uncertainty has affected both foreign direct investment and tourism impacting rates of poverty and unemployment. In September 2011, the national average poverty rate was 11.8 percent with significant variation reaching as high as 29 percent in the Center-West of the country. Furthermore the economy is now estimated to have had negative growth of minus 1.8 percent in 2011, which pushed the unemployment rate to 18.9 percent in 2011 and 18.1 percent in 2012.[1] Without meaningful action, climate change will deepen the already significant poverty and unemployment in the country and may unravel the development gains made in recent decades contributing to food insecurity and political instability. In contrast, policy options that address climate change mitigation and adaptation can create opportunities for economic growth and poverty alleviation.

This report assesses climate risks and opportunities and proposes actions. It provides a synthesis of evidence of climate variability and change, impacts, and uncertainties associated with climate change that may affect Tunisia's water, land, agriculture, and coastal zones. The report then provides a detailed analysis of the potential impacts of climate change on food security and gross domestic product (GDP) as well as on local populations looking in particular at seven governorates. The report goes on to discuss possible policy options for reducing human vulnerability and for better adapting to climate variability and change.

The report provides guidance to policy makers in Tunisia in three ways. First, it provides a Framework for Action on Climate Change Adaptation, represented by an adaptation pyramid. Second, it puts forward a typology of policy approaches that are relevant to the region in order to facilitate the formulation of effective policy responses by decision makers. Finally, a matrix is provided, which outlines key policy recommendations. Proposed actions align with the World Bank's 2012 Interim Strategy Note (ISN) for Tunisia, which guides the World Bank investments in Tunisia over the next two years and is focused on three main areas of intervention: (1) sustainable growth and job creation, (2) the promotion of social and economic inclusion by improving access to basic services for underserved communities and improving the efficiency of social safety net programs, and (3) strengthening governance through improved access to public information as the basis for increased social accountability and transparency.

The Climate is Getting Hotter, Dryer, and More Variable

Meteorological records, derived climate indices, and satellite products show that mean annual temperatures rose by about 1.4°C in the twentieth century with the most rapid warming since the 1970s (figure E.1, left panel). Local rates of warming can be greater. For example, in Tunis, temperatures rose by approximately 3°C during the twentieth century.

Annual rainfall totals averaged across the country show large year-to-year variability linked to atmospheric pressure patterns in the Atlantic and Pacific (the North Atlantic Oscillation and El Niño Southern Oscillation respectively) (figure E.1, right panel). This makes detection of rainfall trends very difficult to determine. However, there is evidence that annual rainfall totals have declined by 5 percent per decade in the north since the 1950s.

Trends in extreme events are also problematic to assess using data available at the time of the study, but regional assessments show that the frequency of warm nights has increased, and heavy rainfall events have become more frequent.

Tidal data suggest that sea levels have risen by more than 3 millimeters per year since 1992 when averaged across the Mediterranean, but records show considerable spatial and temporal variability in local rates. Anecdotal ecological evidence suggests that salt-tolerant vegetation is already migrating inland along canals and drains near the Ghar El Melh lagoon and onto coastal lowlands around the Gulf of Gabes. Coastal land areas are also undergoing salinization, and the sea is inundating some archaeological sites.

Figure ES.1 Twentieth-Century Mean Temperatures (Left) and Precipitation (Right), Tunisia

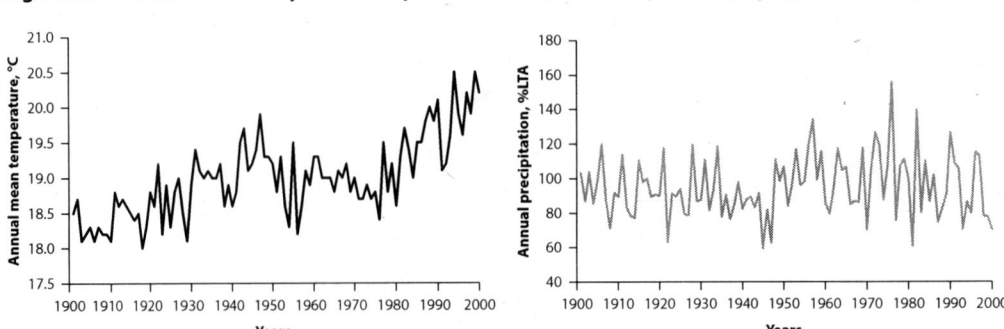

Source: Mitchell et al. 2002; http://www.cru.uea.ac.uk/~timm/data/index-table.html.

Warming and Drying Trends Continue Exacerbating Water Scarcity

By 2050 mean temperatures could change by +1.4°C to +2.5°C and precipitation could reduce by −5 to −15 percent. Local and seasonal changes in temperature and precipitation could be even greater. Future reductions in rainfall are expected to arise from a northward shift in Mediterranean storm tracks.

Sea levels could rise by between 3 centimeters and 61 centimeters during the twenty-first century depending on local water heat content and salinity in the Mediterranean. This could increase the risk of saline water entering coastal aquifers.

Climate Models Can Be Downscaled for Local Climate Change Impact Assessment and Decision making

Two statistical downscaling techniques (SDSM and UCT) were used in chapter 2 of this report to generate illustrative scenarios for Tunis. Daily maximum temperatures (TMAX) and precipitation totals (PRCP) were downscaled from climate model output under the emissions scenario SRES A2. Relative to the UCT downscaling ensemble, SDSM scenarios were found to be at the lower end of warming but more extreme end of regional drying. Depending on the choice of downscaling method and host climate model, the annual mean maximum temperature change could increase by +1.5°C to +2.6°C, and annual precipitation totals could change +5 percent to −20 percent by the 2050s (figure E.2). These scenarios do not reflect the full range of uncertainty because other important drivers of regional climate (such as land surface and aerosol feedbacks on solar radiation) are not included. Therefore, even though most scenarios show warming and drying, this model consensus should not be interpreted as a prediction.

With these warnings in mind, a simple case study demonstrated how downscaled scenarios could be used for climate change impact assessment. Using the

Figure ES.2 Range of Changes in Monthly Mean TMAX (°C) and PRCP (Percent), Tunis

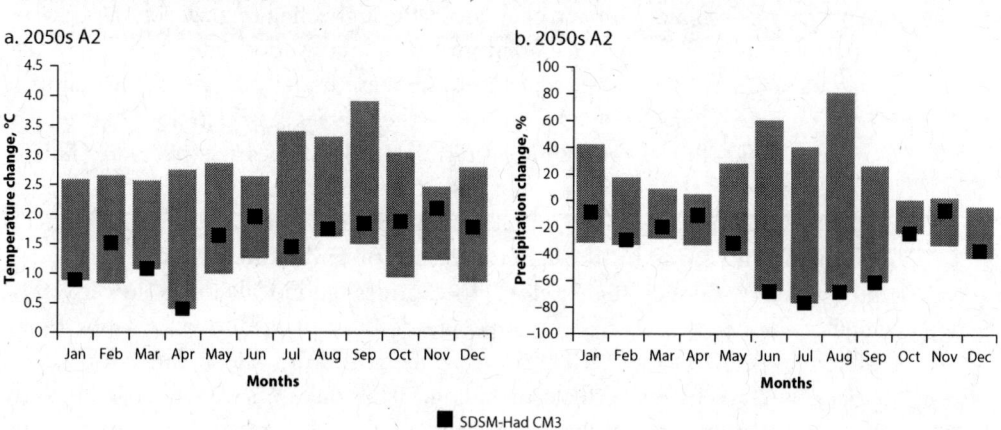

a. 2050s A2

b. 2050s A2

■ SDSM-Had CM3

Source: Based on downscaled by the UCT ensemble (bars) and SDSM (box symbol) under SRES A2 emissions for the 2050s.

case of olive flowering dates, it was shown that rising maximum temperatures in north Tunisia could advance plant growth and flowering by up to three weeks by 2050. Even so, there could still be considerable interannual variability in olive growth and much of the climate response falls within the present range of early/late flowering dates.

The review focused on changes in temperature and precipitation over Tunisia. However, there is evidence that sea level rise is increasing the risk of saline intrusion to coastal aquifers limiting the resources available to irrigators and urban areas on the coast. Other studies suggest that crop yields could be susceptible to changes in growing season and plant moisture availability. Furthermore, higher evaporation demands combined with reduced rainfall could exacerbate soil salinity, and more extreme precipitation could increase erosion on slopes and sedimentation of reservoirs.

The above regional climate change scenarios could accelerate water scarcity and overexploitation of freshwater stocks. However, in the short-and-medium term, population and economic growth are expected to be more important drivers of water stress than climate change. Possible exceptions include situations where a tipping point such as the limit to rainfed agriculture (~200 millimeters per year) or perennial surface drainage (~400 millimeters per year) is being approached. Under even modest rates of climate change these thresholds could be reached with limited time for anticipatory adaptation. Hence, international programs such as CORDEX are seeking to better characterize uncertainty in regional climate projections by prioritizing and comparing downscaling techniques for vulnerable regions including North Africa.

Increased Climate Variability and Change Impact Food Security and GDP

Climate change is expected to have major impacts on Tunisia's agriculture, economy and households from both global and local perspectives. Global climate change's major impact channel is through changing world food prices, especially since Tunisia is a net importer of many food commodities. World market prices for food are projected to increase under climate change and the local climate change impacts manifest themselves through long-term yield changes. Yields for wheat, barley, and irrigated potatoes are expected to fall.

Results from the economic analyses—based on computable general equilibrium (CGE) modeling—show that climate change will lead to negative effects for the overall economy, the agricultural sector, and a total reduction in household incomes. Global (higher global food prices) and local effects (lower yields) together are projected to reduce economic output in Tunisia by US$2.0–2.7 billion over 30 years. Agriculture may benefit from the higher world food prices, but the overall effects of falling yields on the sector are significantly negative. Agricultural growth may drop 0.3–1.1 percentage points by the end of the study period.

Figure ES.3 Impacts of Combined Climate Changes on Household Incomes

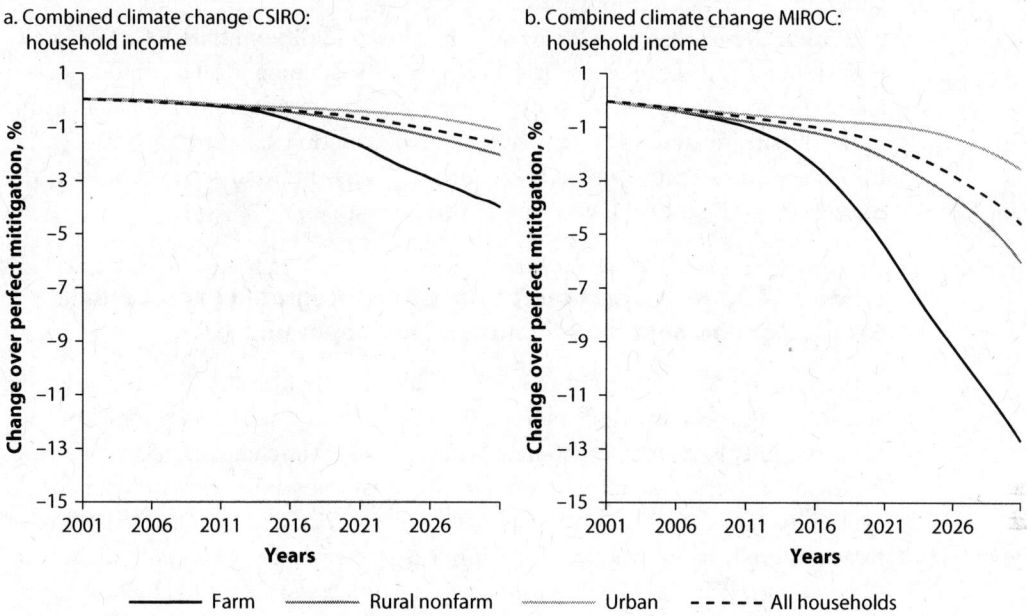

a. Combined climate change CSIRO: household income

b. Combined climate change MIROC: household income

Farm —— Rural nonfarm Urban - - - - All households

Source: World Bank data.

Household well-being is hit even harder. Farm households are the hardest hit by climate change in Tunisia, but rural nonfarm and urban households also suffer (figure E.3). Climate change is projected to reduce farm incomes by 2–7 percent annually, on average over 30 years. Rural nonfarm and urban households as net consumers of food are most affected by the rising global food prices as a result of climate change.

Poor People and Communities Are Among the Most Vulnerable to Climate Change

Climate change will particularly negatively affect the already vulnerable livelihoods of rural populations in the interior regions of central and southern Tunisia, as well as the agro-ecosystems in which they live. These populations earn a significant portion of their income from agriculture, ranging from 13.7 percent of the active population in Tataouine to 30.4 percent in Kasserine, compared to a national average of 16.5 percent. Unemployment rates are also significant. In 2011, unemployment ranged from 12 percent in Kébili to 21 percent in Gafsa, compared to a national average of 16.4 percent. Rates are particularly high among women, ranging from 19.7 percent in Gabes to 28.3 percent in Gafsa. High rates of unemployment have already been contributing to rural to urban migration as well as social unrest. Such problems are likely to increase with climate change. For this reason comprehensive studies were untaken in seven governorates of Tunisia focused on the social implications of climate variability and change for these populations.

Within these areas impacts are multifaceted and differentiated. Food production systems and the agroecological conditions sustaining local livelihood strategies are severely stressed. Historical overexploitation of water and soil, coupled with climate-related stress on living and production conditions, has negative impacts on rural income generation and employment; food security, at both the household and national levels; and natural resources.

Climate Change Adaptation Should Be an Integrated Part of Public Sector Management for Sustainable Development

Governments, sector institutions, and local actors will need to take aggressive action to address climate change in Tunisia. This will not be easy. The Tunisian Government is currently maneuvering in an ever-changing postrevolution political and institutional environment; it must cope with climate change and variability, and depletion of crucial natural resources; and it must deal with the financial crisis in Europe, which Tunisia depends on for exports and tourism. Given the new political environment in the country and the heightened expectations of local communities, there is a need for a systematic, general focus on "low-hanging fruits" or feasible, low-cost adaptive options and there will be a strong need for international cooperation and support to climate change action. This chapter aims to provide clear policy direction and feasible adaptation actions that can help Tunisia address climate change while contributing to economic growth and stability.

The adaptation pyramid (figure E.4, further elaborated in chapter 1) may assist stakeholders in Tunisia in systematically integrating climate risks and

Figure ES.4 Framework for Action on Climate Change Adaptation: Adaptation Pyramid

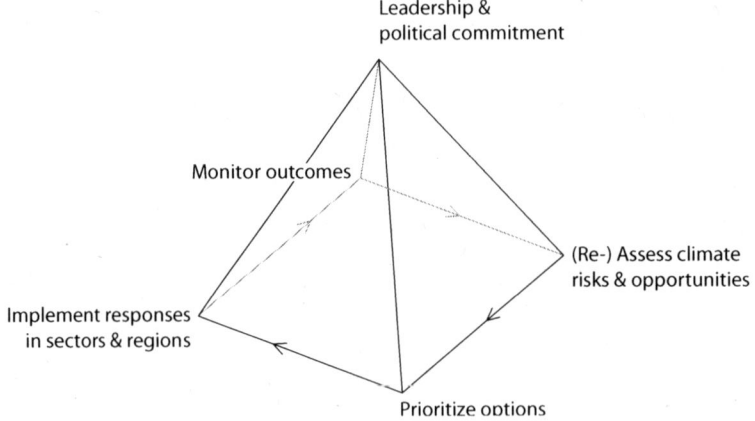

Source: Verner 2012.

opportunities into development activities. It is based on an adaptive management approach critical in the current context of uncertainty. It also highlights the importance of leadership, without which adaptation efforts are unlikely to achieve the necessary commitment to be successful. The base of the pyramid is elaborated in the next sections.

There are four steps in the climate change pyramid:

Assess climate risks, impacts, and opportunities. In this first step, a wide range of analyses could be used.[2] All of these rely on access to climate and socioeconomic data to provide information on climate impacts, including on vulnerable groups, regions, and sectors. To help understand the risks and impacts, data are needed on current climate variability and change as well as projections and uncertainty about the future climate. Similarly, information on past adaptation actions and coping strategies needs to be gathered and evaluated in light of the changing climate.

Prioritize options. The second step is to identify and prioritize adaptation options within the context of national, regional and local priorities. Of particular importance is the identification of financial and capacity constraints. Within this context it is important to consider longer-term consequences of decisions, as short-term responses may not be efficient or could lead to maladaptive outcomes. An example of maladaptation is the further building of housing settlements in highly vulnerable locations. An effective approach for prioritizing options is robust decision making, which seeks to identify choices that lead to acceptable outcomes under many feasible scenarios of the future.

Implement responses in sectors and regions. Adaptive responses will often be somewhat at odds with immediate, local priorities, and thus the third step of implementation requires cooperation and understanding at national, sectoral, and regional/local levels (often jointly). At the national level, adaptation needs to be integrated into national policies, plans, and programs and financial management systems. Such integration includes five-year plans that were prepared in a number of Arab countries. Adaptation planning would need to be integrated into sustainable development and poverty reduction strategies; policies, regulations, and legislation; investment programs; and the budget. Alternatively, national adaptation strategies can guide the mainstreaming of adaptation into other national policies and the implementation of plans at sectoral and local levels.

Monitor outcomes. The fourth step is to monitor outcomes to ensure adaptation-related strategies and activities have the intended adaptation outcomes and benefits. Comprehensive qualitative and quantitative indicators can help planners recognize the relative strengths and weaknesses of various initiatives, which feeds into the next step (reassessing climate risks, impacts, and opportunities) of adjusting activities to best meet current and future needs. The monitoring framework should explicitly consider the effects of future climate change, particularly for projects with long-term time horizons. This is an iterative process.

Effective climate change adaptation will not occur without strong leadership and political commitment. Experience shows that the lead needs to be taken at the national level by a prominent ministry or senior government champion, such as the Prime Minister, Minister of Planning or Economy, or Minister of the Environment. This champion will also require the support of a strong team comprised of representatives of relevant ministries, governorates, local authorities and institutions, the private sector, academia, civil society organizations, and ideally from opposition parties in order to ensure continuity as governments may change. In Tunisia, the support of the Prime Minister, as well as other key ministers such as the Minister of Planning or Economy, or Minister of Environment, will be critical to the effective prioritization and implementation of adaptation actions. Clearly, this should be adapted to the context of individual countries and its circumstances.

Adaptation Decision Making Must Be Supported by a Range of Policy Measures

This report puts forward options that are relevant to the country taking into account the implications of the revolution of January 14, 2011, on the systems of government and policy priorities as well as on the need to prioritize economic growth and employment creation. These policy options relate to five main priority areas:

Improve the Quality and Accessibility of Public Information on Climate Change

Access to quality weather and climate data is essential for adaptation policy making in Tunisia. Without reliable data on temperature and precipitation levels, it is difficult to assess the current climate and make reliable weather forecasts and climate predictions. Tunisia has regular climate data collection with a high concentration of weather stations in the northeast of the country as well as in Djerba and other highly developed regions. There are however, low concentrations of weather stations in the southwest and interior of the country. In the short- and medium term, the collection and monitoring of climate data could be improved by expanding the number of weather stations, and by collaborating with other countries in the region to improve the coverage and comparability of data. Some of these efforts have already begun through Tunisia's valuable participation in the European Climate Assessment and Dataset project, which collates daily observations, performs quality control and analysis, and disseminates the results.

A great deal more work needs to be undertaken in Tunisia to link climate data with water, food and socioeconomic datasets. While information on hydrology is included in the Climdata system used by Tunisia, in many parts of the country, availability of data is poor and will need to be upgraded. In rural areas, it is important to collect data related to changes in agricultural yields and production for indicator crops, forage, and livestock in order to understand the main food supply chains and understand how they might be impacted by climate change. The socioeconomic data types needed for effective policy making include household and census data, and other economic data related to the labor market and production. It is important that social and economic information can be disaggregated to reflect location, gender, age, and socioeconomic status as these factors greatly affect exposure to, and the ability to cope with, climate risks. Data collection methods should be designed to be repeatable, so trends can be tracked closely over time.

Provide Human, Technical and Other Resources and Services to Support Adaptation

Specialized human and technical resources are required to analyze, identify, and implement adaptive responses. Human and technical resources can be developed through education and training, research and development, and technical improvements. Education has been a priority of the GoT and the country's education system has ranked as one of the best in the Middle East and North Africa in the Human Development Index as well as in Organisation for Economic Co-operation and Development (OECD) and United Nations Development Programme (UNDP) studies. Environmental issues have been incorporated into school curriculums through the Sustainable Schools Program.[3] However, climate change adaptation is not yet incorporated into this program. In several governorates visited for this report, individuals reported a growing awareness of a changing climate. There was, however, a limited understanding of the linkages between this knowledge and effective climate change adaptation.

A number of education institutes and new environmental education programs in primary and secondary schools already include some climate training. At a graduate level, the University of Tunis offers courses in climatology and climate change. The National Institute of Agronomy includes training in agroclimate linkages. More could be done to integrate climate change policy into university courses and to create linkages between the technical courses on climatology and general policy making. Specific groups of individuals should also be targeted for education and training. Midcareer professionals engaged in particular sectors would be one such group. For example, training for water utility employees to enhance water demand management through market based instruments (for example, water pricing, metering). Rural women who are coping with the effects of out-migration of men could be trained in community and political participation skills, business development, general literacy, and education and extension services. Other groups to target include local government officials, emergency management teams, and poor people engaged in climate-exposed livelihoods. Training can also be targeted to particular governorates that are subject to high climate risk.

Tunisia's national government also has an important role to play in facilitating promotion of and access to technologies that help people to adapt to climate risks. This is best accomplished through a combination of policy reforms and financial interventions that change incentives for private investment in new technology and address key market failures. In the water sector, priority areas include reducing water network leakage, improved storage and conveyance capacity and reducing water demand through, for example, drip irrigation. New types of food storage or food transport systems could also be supported to ensure food security and improve the transportation of agricultural goods to market. Technology transfer is sometimes available but often these technologies can be derived only through local research and development. Recent agreements with India to promote cooperation in biotechnology research offer some prospects for future growth in Tunisia's national technology development related to climate change, as do efforts under way to address water shortages through desalinization technology.

Build Climate Resilience Through Social Protection and Other Measures

The second priority of the World Bank's 2012 ISN for Tunisia is the promotion of social and economic inclusion by improving access to basic services for underserved communities and improving the efficiency of social safety net programs. Efforts to promote social protection including insurance schemes, pensions, access to credit, cash transfer programs, and relocation programs are critical aspect of building resilience to climate change.

Underinvestment in social safety nets, public services such as water supply and wastewater treatment, and housing and infrastructure, as well as inequitable access to health care make people more vulnerable to a changing climate. In response to the revolution of January 2011, the Government of Tunisia is seeking to reform its social protection system with a study of social vulnerability currently under way. At present, however, climate change has not been considered as a key factor for social vulnerability in this study. In the context of increasing climate change impacts, which particularly affect the poor and vulnerable in many regions of the interior, it will be essential to incorporate climate change into these studies.

In rural areas with high rates of out-migration of men, social protection is particularly critical for the women, elderly, and children left behind. This can take the form of rural pension schemes or conditional cash transfer programs. Assistance to enhance productivity can include the facilitated access credit or markets for agricultural and other rural products. In urban areas, social services can include the provision of affordable housing away from locations at risk of climate impacts such as flood zones or the provision of energy, water, and public transport services. The poor and most vulnerable are particularly in need of assistance when an extreme weather event hits. This could include emergency response provisions of basic needs, like adequate shelter and access to food, water, and clothing.

Develop a Supportive Policy and Institutional Framework for Adaptation

A supportive policy and institutional framework at national, sectoral, and local levels is essential for effective climate change adaptation decision making. Basic conditions for effective development such as the rule of law, transparency and accountability, participatory decision-making structures, and reliable public service delivery that meets international quality standards are conducive to effective development and adaptation action. In addition, climate change adaptation requires new or revised climate-smart policies and structures at all levels. A clear but coordinated governance structure is also essential to implement climate change adaptation measures.

Tunisia is increasingly showing progress integrating environment and climate change into national policies. Tunisia has already developed a number of national adaptation strategies as well as sectoral strategies such as the strategy on the adaptation of agriculture and ecosystems to climate change (January 2007), the strategy on the adaptation of the coastal zones to climate change (February 2008), and the strategy on the adaptation of the public health sector to climate change (2010). Throughout 2011 and 2012, a process has been under way in cooperation with the GIZ, the World Bank and other donors to develop this National Strategy for Climate Change. On May 23, 2012, Tunisia signed up to the OECD Declaration on Green Growth, and has been actively engaged in the Green Growth strategy development following Rio+20 and in low-carbon and climate resilient development activities showcased at the climate negotiations in Doha. The implementation of national and sectoral climate change strategies within Tunisia will require interministerial coordination as well as strong coordination between different levels of government. This is an area in need of enhanced attention and support in Tunisia.

Regional and international collaboration is also essential for climate resilience. The heterogeneity of the Arab countries and diverse national experiences provides multiple opportunities for beneficial climate related regional collaboration. Tunisia will be best equipped to address climate change if they have strong collaboration on issues such as climate-related data sharing, crisis response, the management of disease outbreaks, migration, shared water resources, and strong trade relationships to address food security. Engaging with international bodies can link Tunisia to new initiatives and knowledge and prove key in improving climate change responses. For example, the World Meteorological Organization is promoting a new large-scale initiative on climate services that could benefit from enhanced participation from Arab States.

Build Capacity to Generate and Manage Revenue and to Analyze Financial Needs and Opportunities

Financial resources are essential for development and to effectively adapt to climate change. Tunisia will need to invest in building the capacity to analyze their financial needs and generate and manage climate change related resources. Ministries need to mainstream climate change adaptation into national and sectoral budgets by developing systems to track climate change finance in domestic budgets and allocating finance to activities that reduce vulnerability to climate change. Moreover, current and future climate change impacts need to be taken into account in planning and costing investments, particularly over the long term. Financial resources for climate change will need to come from domestic and international sources. For Tunisia, national public expenditure reviews could be one tool to highlight current expenditures and better understand how these relate to budget estimates for climate proofing infrastructure. This information, in turn, will help Tunisia to understand what levels of additional revenues are needed to make up for shortfalls and identify new revenue opportunities. These revenues could come from the removal of subsidies, creation of innovative tax mechanisms, or Payments for Ecosystem Services (PES). PES has significant potential to enhance rural livelihoods and agricultural yields, maintain and enhance ecosystem services such as watersheds and biodiversity, and develop long-term partnerships with the private sector. PES can contribute to disaster risk reduction, with the revenues generated serving as financial buffers for communities to climate-induced shocks.

There Are a Number of Priority Investments for Tunisia

Invest in agricultural research and development, particularly in the breeding of climate-proof crops. Yields of rainfed crops in Tunisia are hit especially hard. Scientific advancement for breeding more climate-proof varieties will therefore be key for the future of agriculture in Tunisia and the Arab region more broadly. Farmers also have different on-farm management techniques to offset the impacts of climate change which may include: Shifting the planting date, switching crop varieties, switching crops, as well as expanding the area of production and/or increasing irrigation coverage (Burke and Lobell 2010) and improving irrigation efficiency especially in the face of already-constrained water resources. Furthermore, research and development in agriculture would also include changes in crop practices—optimum sowing dates, choice of cultivars, and planned plant density (Hainoun 2008)—re-evaluating and redesigning irrigation, and water harvesting practices to sustain a healthy agricultural sector. In trying to address climate change, it is essential to distinguish between short-term measures and long-term measures.

Encourage and support risk management strategies. It is crucial to build a rich and functioning network for risk mitigation that include social and extension services linking farmers to agricultural research as well as linking the vulnerable population to markets and policy makers. A network of communication and extension services is crucial in outreaching to farmers, or the agricultural community as a whole. Such a comprehensive system ensures the dissemination of relevant information, and/or techniques and cultivars and guarantees that national policies are implemented down to the individual unit; the farmer.

Furthermore, such a network also provides a strong link back from the farmer to scientists and policy makers for the collection of information relevant for technological advance and policy making. Disaster risk management strategies, such as index-based weather insurance schemes can be a powerful tool to mitigate the risk facing small farmers' livelihoods due to weather variability and consequent crop loss. The advantages of simple weather security schemes are several. Insurance would be provided through groups to reduce the transaction costs for the insurance company (Martins-Filho et al. 2010), they would increase coverage on weather variability to small farmers, which translates into less livelihood disturbances and risk. In order to successfully operate, there has to be in place a relevant weather index which the insurance schemes are tied to, in order to sustainably provide timely and accurate information. Also, given the reliance on the group insurance structure for these schemes, there needs to be in place, or under construction, strong farmer extension channels for product and information dissemination.

Foster job creating growth and social protection schemes to support the rural and urban poor. Good development policy is good climate change adaptation strategy for the rural and urban poor. Even if the severity and frequency of variable climates remains constant, the impacts are likely to have increasingly negative socioeconomic consequences as a result of a larger population (thus increasing demand for food) and increasing groundwater depletion. The rural farm households and the rural nonfarm and urban poor are particularly hard hit, mainly due to the large share they spend on food out of their income and the more reliant rural farm households are on farm income for their livelihoods. It is essential to have in place social safety nets that will provide the necessary channels of outreach and mitigation to the poor and vulnerable, both, in times of crisis and under more benign conditions, to disseminate product information and technical support. In general, social safety nets and long-term development goals should be integrated together as well as with national goals and objectives in Tunisia's national plan (see table ES.1).

Develop a comprehensive food security strategy to prepare Tunisia for rising and more volatile global food prices. Given that climate change translates into higher and more volatile global future food prices, and Tunisia is likely to become more dependent on food imports in the future, a forward-looking national food security strategy is urgently needed. Important components of such a strategy should comprise (1) assessing the potential and future role of agriculture for the economy and for food security; (2) future allocation of water between agriculture, commercial, and household use; (3) revisiting international trade agreements and domestic food supply chains; (4) health and policies related to population growth, poverty reduction, and nutrition; and (5) food security risk management. On the global level, Tunisia may actively engage in ongoing effort to reform the global food system to ensure a less volatile and more food secure future for all.

Table ES.1 Policy Matrix

	Collect information on climate change adaptation and make it available	Provide human and technical resources and services to support adaptation	Provide assistance such as social protection for the poor and most vulnerable	Ensure a supportive policy and institutional framework	Build capacity to generate and manage finance and analyze financial needs and opportunities
General	– Establish a central bureau holding and disseminating data sets to concerned sectors on observed climate, socioeconomic characteristics, land use, and climate change scenarios.	– Identify needs (human and material). – Reinforce capacity for regional studies at the central level. – Put in place a process for regional consultations implicating nongovernmental organizations (NGOs). – Mainstream climate change adaptation into regional planning. – Put in place a system of monitoring and evaluation.	– Establish programs to support the basic needs and employment of the poor and most vulnerable.	– Support the implementation of the national strategy for adaptation. – Mainstream climate change considerations into the new constitution.	– Ensure mainstreaming of climate change at all levels and in all stages of planning and budgeting. – Put in place an institution responsible for coordinating access to external finance for climate change adaptation.
Climatology	– Climate scenario and impact analyses with other countries in the region (recognizing that many climate risks transcend state boundaries). – A MENA workshop to share lessons learnt and findings from climate studies across the Arab region as a whole.	– Enhance national and/or regional capacity to utilize existing international programs on satellite retrievals and data. – Train and enhance capacity to work with and use comprehensive data sets such as reanalysis products. – Build capacity to use regional climate data information[a]	– Empower civil authority to be in charge of making data available for public use.	– Enhance regional collaboration on early warming systems, including use and dissemination of existing extended forecasts (available through WMO, etc.).	– Include climatological data collection in the national budget including costs related to data rescue; extending the number of weather stations.

(table continues on next page)

Table ES.1 Policy Matrix (continued)

	Collect information on climate change adaptation and make it available	*Provide human and technical resources and services to support adaptation*	*Provide assistance such as social protection for the poor and most vulnerable*	*Ensure a supportive policy and institutional framework*	*Build capacity to generate and manage finance and analyze financial needs and opportunities*
	– Collaborative research projects involving national and international experts addressing specific knowledge gaps: (1) coproduction and validation of climate scenarios for Tunisia; (2) enhanced capability for seasonal forecasting of drought at national levels and for agro-economic regions [for example, Normalized Difference Vegetation Index (NDVI), Standardized Precipitation Index (SPI), food price indices]; and (3) building technical capacity for impacts modeling. – Extend the coverage of the observational network in order to ensure a minimal station density to reflect climate variability and likely change in the country/region (also beneficial to weather forecasting and early warning systems).	– Promote and use available products for impacts and climate change risks analyses among users. – Establish regional/international centers of excellence. – Improve use of knowledge of centers existing within the country through staff exchange and through enhanced regional and international cooperation.	– Combine climate data with socio-economic data in order to obtain information that can assist in building resilience, including for the poor and most vulnerable.	– World Meteorological Organization (WMO) is promoting a new large-scale initiative on climate services. This initiative is crucially dependent on the active participation of the member states; capacities to explore and contribute to these efforts are critically needed. – Encourage international collaboration—wealth of satellite information can complement ground information	– Establish centers of excellence. – Build capacity and training.
Rural	– Assess changes in agricultural production levels/yields for indicator crops. – Model the food supply chains, model how they operate, and how they will be impacted by climate change.	– Develop knowledge and skills related to climate resilient agricultural practices such as growing salt-tolerant, heat-tolerant, and pest-resistant crop and livestock species, conservation agriculture, increasing irrigation efficiency and using nonconventional water resources.	– Use targeted transfers during price spikes and crop failures to support the most vulnerable. – Support development of access to markets for agricultural and other rural produce.	– Create a clear but coordinated governance structure to implement climate change adaptation measures at central and local levels across ministries responsible for agriculture, water, and the economy.	– Develop capacity to estimate the financial risks for not applying climate change adaptation and how to maximize risk management through available financial instruments.

(table continues on next page)

Table ES.1 Policy Matrix *(continued)*

Collect information on climate change adaptation and make it available	*Provide human and technical resources and services to support adaptation*	*Provide assistance such as social protection for the poor and most vulnerable*	*Ensure a supportive policy and institutional framework*	*Build capacity to generate and manage finance and analyze financial needs and opportunities*
– Monitor state of water (groundwater and salinity levels) and soil conditions (depth and carbon content), and agricultural activities in "most at risk" agricultural zones (using indicator areas of marginal lands, rainfed areas from the four regions).	– Develop human and technical resources to optimize food chain systems particularly in transport, marketing, improving value-added developments, and establishing cooperatives.	– Support development of schools and training facilities to nurture both basic academic and vocational skills and provide necessary incentives to ensure attendance is possible.	– Develop a coordinated national policy, likely to be across ministries, supporting food security and rural livelihood developments, balancing risks with possibilities and mindful of water and energy security vulnerabilities. – Create farmers' associations that link directly to the Ministry of Agriculture and agricultural research/extension services to ensure clear flow of knowledge to all areas from top-down and bottom-up.	– Enhance capacity to assess all possibilities for meeting food demand balancing economics with geopolitical risks.

Source: World Bank data.

Note: MENA = Middle East and North Africa; NGOs = nongovernmental organizations; WMO = World Meteorological Organization.

a. For example by means of Geographical Information Systems (GIS).

Notes

1. Based on figures from the National Statistics Institute of Tunisia accessed on November 12, 2012 at http://www.ins.nat.tn/indexen.php see report on unemployment http://www.ins.nat.tn/communiques/Note_emploi_1T2012_15052012_V3.pdf.

2. For an overview of available tools to assist in climate risk analysis see http://climat-echange.worldbank.org/climatechange/content/note-3-using-climate-risk-screening-tools-assess-climate-risks-development-projects.

3. *Programme des écoles durable.*

References

Burke, M., and D. Lobell. 2010. "Food Security and Adaptation to Climate Change: What Do We Know?" In *Climate Change and Food Security*, edited by D. Lobell and M. Burke, 133–53. Dordrecht, The Netherlands: Springer Science + Business Media, B.V.

Hainoun, A. 2008. "Vulnerability Assessment and Possible Adaptation Measures of Agricultural Sector." United Nations Development Programme, Unpublished Report.

Martins-Filho, C., A. S. Taffesse, S. Dercon, and R. V. Hill. 2010. *Insuring Against the Weather: Integrating Generic Weather Index Products with Group-based Savings and Loans.* Seed Project Selected. United States Agency for International Development: Washington, DC. http://i4.ucdavis.edu/projects/seed%20grants/MartinsFilho-Bangladesh/files/Martins-Filho%20proposal%20short.pdf.

Mitchell, T. D., Hulme, M. and New, M. 2002. Climate data for political areas. Area, 34, 109–112. http://www.cru.uea.ac.uk/~timm/data/index-table.html

Verner, D., ed. 2012. "Adaptation to a Changing Climate in the Arab Countries: A Case for Adaptation Governance and Leadership in Building Climate Resilience." MENA Development Report, World Bank, Washington, DC.

CHAPTER 1

Climate Change Is Happening and People Are Affected

Photograph by Dorte Verner

In Tunisia and across the globe, climate change is already damaging people's livelihoods and well-being. It is a threat to poverty reduction and economic growth and may unravel many of the development gains made in recent decades. Both now and in the long run, climate variability and change threaten development by restricting the fulfillment of human potential and disempowering people and communities, constraining their ability to protect and enrich their livelihoods. It is therefore essential to invest in climate change adaptation. This involves adopting measures to protect natural and human systems against the actual and expected harmful effects of climate change; to exploit any opportunities climate change may generate; and to ensure the sustainability of investments and development interventions in more difficult climatic conditions.

A harsh climate has shaped the culture in many parts of Tunisia for thousands of years. Over the centuries the people of Tunisia have coped with the challenges of climate variability by adapting their survival strategies to changes in rainfall and temperature. However, the message in chapter 2 is clear: over the next century this variability will increase and the climate will experience unprecedented extremes. Many climate adaptation strategies that people have utilized throughout history have become less viable. For example, the only choice left for people in drought-stricken rural areas faced with depleted assets and reduced productivity is to move to cities or towns where their rural skill set is hard to deploy (Verner, forthcoming).

Existing climate-related challenges in Tunisia include water scarcity, very low and variable precipitation, and exposure to extreme events including drought and floods. If no drastic measures are taken to reduce the impacts of climate change particularly in the agriculture and water sectors, the country will be exposed to increasingly reduced agricultural productivity and incomes, due to a higher likelihood of drought and heat waves, a long-term reduction in water supply, and the loss of low-lying coastal areas through potential salt water intrusion, and eventually to sea level rise. This exposure to climate impacts will have considerable implications for human settlements and socioeconomic systems (IPCC 2007). As climate variability increases so does vulnerability to it, especially for the poor and those heavily dependent on natural resources such as the farmers and pastoralists (see chapter 4).[1]

Climate change has or will soon affect most of the 10.6 million people in Tunisia. The roughly 3.5 million people in the rural areas are likely to be among the hardest hit, as they tend to have the least resources at their disposal to cope and successfully adapt. Therefore, they are likely to be the hardest hit by climate variability and change.[2]

Climate change adaptation can reduce the poor's vulnerability to negative climate impacts. Adaptation measures include the development and implementation of innovative technologies, diversification and adoption of alternative livelihoods, and provision of social protection schemes (see chapter 5). It is also important to note that the poor contributed the least to Tunisia's carbon dioxide (CO_2) emissions of 2.4 metric tons per capita, which is less than 0.1 percent of global CO_2 emissions.

There is increasing evidence that climate change will have severe negative impacts on the economic and social development of Tunisia. Climate change threatens to stall and reverse progress toward poverty reduction, improved health, gender equality, and social inclusion.[3] Yet, research on the socioeconomic dimensions of climate change in the Arab region is only in its early stages (Tolba and Saab 2009). This report aims to assess the impacts of climate variability and change in order to fill knowledge gaps and respond to the Government of Tunisia's (GoT) request for technical assistance in understanding and identifying successful adaptation policies and programs. This analysis will support the country and its people in building resilience to climate change, particularly for the poor and vulnerable (see box 1.1 for the IPCC definitions related to climate change).

This report serves as a resource to begin to assess climate risks, opportunities, and actions. The report explains the potential impacts of climate change in Tunisia and then goes on to discuss possible policy options to reduce climate risk and better adapt to climate variability and change. The methodology applied combines quantitative and qualitative approaches to assessing climate change impacts and adaptation behavior. This report consists of five chapters.[4] This chapter presents the motivation and instruments to address climate change in Tunisia. Climate change scenarios and impacts on Tunisia are addressed in chapter 2. Economic impacts of climate change on Tunisia are addressed in chapter 3 and socioeconomic impacts of climate change in Central and Southern

---

Box 1.1

IPCC Definitions: Climate, Climate Change, and Climate Variability

Climate in a narrow sense is usually defined as the average weather, or more rigorously, as the statistical description in terms of the mean and variability of relevant quantities over a period of time ranging from months to thousands or millions of years. The classical period is 30 years, as defined by the World Meteorological Organization (WMO). These quantities are most often surface variables such as temperature, precipitation, and wind. Climate in a wider sense is the state, including a statistical description, of the climate system.

Climate change refers to a statistically significant variation in either the mean state of the climate or in its variability, persisting for an extended period (typically decades or longer). Climate change may be due to natural internal processes or external forcings, or to persistent anthropogenic changes in the composition of the atmosphere or in land use.

Climate variability refers to variations in the mean state and other statistics (such as standard deviations, the occurrence of extremes, and so on) of the climate on all temporal and spatial scales beyond that of individual weather events. Variability may be due to natural internal processes within the climate system (internal variability), or to variations in natural or anthropogenic external forcing (external variability).

Source: Glossary, IPCC. 2001.

Tunisia in chapter 4. Finally, this report provides guidance to policy makers. It does this in three ways. First, it provides a Framework for Action on Climate Change Adaptation, represented by an Adaptation Pyramid, which aims to provide an illustration of the key elements of iterative adaptation decision making (see figure 1.2). Second, it puts forward a typology of policy approaches that are relevant to Tunisia, to support adaptation decision making. Third, it provides a policy matrix which outlines key policy options covered in each of the areas of this report. These policy recommendations and options were developed together with the GoT, led by the Ministry of Environment, and with planning, meteorological, and regional organizations and institutes, and other stakeholders.

In this context, the GoT is working to develop national studies, strategies, action plans, and project portfolios for key sectors with respect to climate change mitigation and adaptation (see box 1.2). The World Bank plays a complementary role by providing technical assistance through this task, and by supporting investments in sectors such as agriculture, energy, health, and water, among others. This report is responsive to the objectives of the World Bank fiscal 2013/14 Interim Strategy for Tunisia, in its three areas of engagement: (1) laying the foundation for sustainable growth and job creation; (2) promoting social and economic inclusion; and (3) strengthening governance: voice, transparency, and accountability. This study complements the sectoral assessments prepared by the Government of Tunisia by adding overall economic, social and regional analyses.[5]

Box 1.2

Ongoing Work in Tunisia on Climate Change

To address climate change impacts, the GoT (Ministry of Environment) is in the process of completing a National Climate Change Strategy, building on and updating previous work. In July 2012, a two-day climate scenarios workshop was held at the National Meteorological Institute to provide a primer on latest climate change science and practical experience of using climate scenario tools. This is part of wider initiatives to build technical capacities in climate scenario generation and risk assessment.

Existing analyses include a national strategy and action plan for climate change adaptation for the agricultural sector and agro-ecosystems. The action plan is centered on three principal axes: (1) overcoming short term crisis management through a risk adaptation strategy linked to climate change, (2) integrating climatic volatility within agricultural and economic policies, and (3) managing the socio-economic consequences set to impact the agricultural sector in an integrated manner between economic sectors.

A national strategy and action plan have also been completed for adaptation to sea level rise in coastal areas and ecosystems focusing on five principal themes: (1) oceanic and sea level monitoring through the creation of the Sea Level Observatory (ONMER),[6] (2) adaptation

(box continues on next page)

Box 1.2 Ongoing Work in Tunisia on Climate Change *(continued)*

measures for coastal areas, (3) water resources, including measures for vulnerable coastal aquifers, (4) ecology and fishery resources, and (5) coastal infrastructure, including measures for port and sanitation system infrastructure.

Other technical ministries have also developed adaptation strategies and corresponding action plans for the health and tourism sectors. The agriculture sector adaptation strategy is itself complemented by three studies on early warning systems to manage the risk of extreme events. The tourism adaptation strategy is complemented by three studies on the development of ecotourism. These sector strategies and actions plans will be integrated into the forthcoming National Climate Change Strategy mentioned above and a multi-sectoral adaptation project portfolio. Project categories under the portfolio will respond to the following priority areas: (1) water resources, (2) agriculture, (3) biodiversity and ecosystems, (4) industry and energy, (5) wastes, and (6) health.

A number of mitigation activities are also already under way. The government has completed a portfolio of Clean Development Mechanism (CDM) Projects. A national strategy complements the portfolio by focusing on the acceleration of the CDM process across sectors and the implementation of programs for capacity building, technical support, financing, and knowledge dissemination for private or public projects. Thus far, the emission reduction potential is measured at 12.7 million tCO_2e (tons of carbon dioxide equivalent) for 2006–11 and approximately 17 million tCO_2e for 2012–16, with 74 projects covering the areas of energy, wastes, forestry, and industrial processes.

The GoT has positioned itself to access various climate funds for mitigation activities. Through the World Bank, the Government accesses the Clean Technology Fund (CTF) for the Concentrated Solar Power (CSP) program, and carbon financing for the Sidi Daoud wind farm project. The GoT also plans to access the Energy Sector Management Assistance Program (ESMAP) for the Regional Energy Management Incentive (REMIT) project, GEF funds for expanded investments in energy efficiency, and possibly a Multiple Donor Trust Fund (MDTF) for climate change.

Source: Based on Ministry of Environment/GIZ (2007, 2009, 2011), UNDP (2011), and World Bank (2011).

Climate Change Is Happening Now

Climate change is already being felt in Tunisia. A snapshot of the scientific and media reports on climate change in Tunisia shows the country's increasing profile (see chapter 2 for detailed description as it provides the first substantive synthesis of climate change information for Tunisia since the First National Communication to United Nations Framework Convention on Climate Change [UNFCCC] in 2001):

- Higher temperatures and more frequent and intense heat waves threaten lives, crops, terrestrial biodiversity, and marine ecosystems.
- Reduced but more intense rainfall cause both more droughts and more frequent flash flooding.

- Loss of winter precipitation storage in snowmass induces summer droughts.
- Increased frequency of prolonged droughts leads to losses in livelihoods, incomes, and human well-being
- Sea-level rise threatens coastal cities, wetlands, and small islands with storm surges, saltwater intrusion, flooding, with associated human impacts.
- Changing rainfall patterns and temperatures create new areas exposed to vector and waterborne diseases affecting people's health and productivity.

Climate Change Impacts Are Socially Differentiated

Climate change affects all people in Tunisia. Still, the effects of climate change are regionally and socially unequal. Asset-poor communities have few resources but some capacity to adapt to the changing climate. Many manage to take actions by diversifying their livelihood, moving, pursuing education, and so on (see box 1.3). Climate change is superimposed over the preexisting risks and vulnerabilities that poor and marginalized groups typically face. Many studies have suggested that the poor are the most vulnerable to climate change because of their

- Dependence on natural resources, which are exposed to climate change impacts
- Lack of assets, which hinders effective adaptation
- Settlement in at-risk areas, which are less productive and also vulnerable to floods or drought, or other severe events
- Migrant status, which can prevent them from accessing certain social services
- Low levels of education, which prevents them from developing more climate-resilient skills or livelihood strategies
- Minority status, which deters policy makers from making them the focus of adaptation policies.

The Tunisian people increasingly do not know what to expect regarding the climate, and hence what decisions to take. This is especially the case for climate-dependent activities such as agriculture, given changes in the timing and intensity of rainfall and the variability in temperature. A key asset of farmers is their traditional knowledge of the environment, but this knowledge may no longer be reliable without the support of forecasting technology and additional climate information.

The impacts of climate change vary across regions in Tunisia. The approaches to address these impacts depend in part on the *perception* of climate change impacts and in part on the *capacity* to forge responses to them. This report provides strategic guidance on adaptation with a short-time horizon—until 2030—while taking into account longer-term climate change projections.

Many poor people are already being forced to cope with the impacts of climate change. Small farmers are experiencing reduced crop yields and lost outputs due to climate variability and change. Unemployment is increasing, particularly in sectors and regions vulnerable to climate change impacts.

Climate change is a threat to short-, medium- and long-term development. It restricts human potential and reduces the ability of people to make informed choices regarding their well-being and livelihoods (Mearns and Andrew 2009; Verner 2012). As the Stern Review (2007) argues, it is paramount for climate change issues to become fully integrated into development policy and to increase international support for these measures. Social, economic, and human development is key to the efforts to reduce potential conflicts, migration and displacement, losses to livelihood systems, and damages or declines in infrastructure. Effectively addressing all of these issues will help enable people and communities to cope with climate change.

Box 1.3

Geographic and Social Political Context

Tunisia is a relatively small country (163,610 square kilometers), situated between Algeria and Libya and bordered in the North and East by the Mediterranean Sea. The North is character-ized by green rolling hills and a relatively humid climate receiving 500–1,500 millimeters of rain annually and temperatures ranging between 12°C and 30°C. In contrast, the Centre is semi-arid with only 150–400 millimeters of rain annually. The south is characterized by the sand dunes of the Sahara Desert, and annual precipitation does not exceed 178 millimeters a year, with maximum temperatures over 50°C in summer.

This diverse geography will result in differential climate change impacts. The South will be subject to the highest increases in annual temperatures and most significant reduction in annual rainfall. The Centre will also experience significant temperature increases and rainfall reduction. The North will be subject to the smallest increases in annual land seasonal tem-peratures and the smallest reduction in rainfall. In addition to these changes, a rise in sea levels will also be witnessed, threatening increased erosion, flooding, and salinization of aqui-fers in coastal areas.

The revolution that launched the Arab Spring has created significant change in Tunisia resulting in new challenges and opportunities for addressing climate change. Open access to information, the right to elect legitimate representatives and institutions that guarantee accountability, a temporary commission to investigate corruption, and the introduction of new laws that will provide the statutory framework for stronger and independent institutions may increase donor confidence. Furthermore, the growing efforts to address the inequality that sparked the revolution may encourage forms of social protection aimed at the poorest and most vulnerable to climate change.

In addition, clearer statistical information on poverty and unemployment may help target climate change responses. In 2005, official reports from Tunisia provided a national poverty rate of 3.8 percent. After the revolution, the National Statistics Institute published revised poverty estimates in September 2011, indicating that the national average poverty rate in 2005 was 11.8 percent. The report produced in September 2011 also showed for the first time

---

(box continues on next page)

Box 1.3 Geographic and Social Political Context *(continued)*

a breakdown of poverty rates by region indicating large variations, with poverty rates as low as 5–7 percent in the Center-East and Grand Tunis region and as high as 29 percent in the Center-West of the country.

In the aftermath of the revolution, political uncertainty has affected both foreign direct investment and tourism, impacting already-high rates of unemployment and slowing economic growth. This has been compounded by the global financial crisis, which has led to a decline in exports. The drop off in demand from the European Union in particular, Tunisia's main trading partner, has had a significant impact. The conflict in Libya produced an influx of refugees, and the return of Tunisian migrant workers. While many of the refugees have now returned to Libya, the loss of remittances from migrant workers is estimated at US$48–83 million. The cumulative result is that the economy is now estimated to have had negative growth of minus 1.8 percent in 2011, which pushed the unemployment rate to 18.9 percent.

Climate Change Impacts People and the Economy

Climate change puts additional stress on people and the economy. Climate variability and change can lead to, and add to, disruptions to social, infrastructural, environmental, or productive systems and resources, which in turn can slow economic growth and increase poverty. Regions that rely heavily on climate-sensitive sectors, such as agriculture, fisheries, and tourism, and have high poverty rates, lower levels of human capital, or less institutional, economic, technical, or financial capacity will be the most vulnerable.

The total population of Tunisia is fairly constant at around 10.6 million; with a slight tendency toward decreasing (see table 1.1). Better social and infrastructural services in cities combined with climate stress have already led to the rapid urbanization of many Arab countries. As a result, millions of people have left their rural homes to settle in urban centers. The most recent data show that 67.3 percent of the people in Tunisia live in urban areas, which are growing at a rate of 1.6 percent a year. Cities have specific vulnerabilities made worse by rapid growth, partly driven by the migration of the rural poor. However, although less than 35 percent of the country's population live in rural areas, 25.8 percent is employed in agriculture, which contributes 8.0 percent to the national annual per capita gross domestic product (GDP) of US$8,566 (see table 1.1). It is therefore necessary that climate change adaptation occur in both rural and urban areas.

People are vulnerable to the impacts of climate variability and change in light of associated changes in water availability, food security and their health. Although different adaptation options will be deployed in these different environments, it is important that in both settings, local women and men, especially the poor, play an integral role in building the resilience of their

Table 1.1 Socioeconomic Information for Selected Arab Countries

	Algeria	Libya	Morocco	Tunisia	Egypt	Jordan	Lebanon
Land area (km^2)	2,381,740	1,759,540	446,300	155,360	995,450	88,780	10,230
Agricultural land (% of land area)	17.4	8.8	67.3	63	3.7	11.5	67.3
Forest land (% of land area)	0.6	0.1	11.5	6.5	0.1	1.1	13.4
Population (millions)	35.5	6.4	32	10.6	81.1	6.1	4.2
Population growth (annual %)	1.5	1.5	1	1	1.8	2.2	0.7
Urban population	66.5	77.9	56.7	67.3	42.8	78.5	87.2
Urban population growth (annual %)	2.4	1.7	1.6	1.6	1.8	2.3	0.9
Urban pop >1 million (% of total)	7.9	17.4	19.3	0	19	18.3	45.8
Population in areas with elevation <5 m	3.5	4.7	3.8	9.5	25.6	4.2	9.1
GDP per capita, PPP (cont. 2005 $)	7,521	15,361	4,227	8,566	5,544	5,157	12,619
Human Development Index (HDI) Value 2011	0.70	0.76	0.58	0.70	0.64	0.70	0.74
Poverty ratio at $1.25 a day	6.8	–	2.5	2.6	2	0.4	–
Labor force, total (1,000s)	14,845	2,305	11,846	3,821	26,536	1,818	1,444
Employ. in agriculture (% total employment)	20.7	19.7	40.9	25.8	31.6	3	–
Agriculture, value added (% of GDP)	11.7	1.9	15.4	8.0	14.0	2.9	6.4

Source: Data from World Bank World Development Indicators (December 2011) and UNDP Human Development Report 2011. Data are the most recently available and most apply to 2010.
Note: ppp = purchasing power product; GDP = gross domestic product.

livelihoods and well-being. Women can be, at the same time, the most vulnerable group, and the main agents for the social changes needed for coping with a changing climate.

It is projected that the Tunisian economy will be increasingly affected by climate change. This is illustrated by this report in the analysis on income, livelihoods, well-being, poverty, and other social factors (see chapters 3 and 4).[7] While experts agree on climatic trends (see chapter 2), it is less clear what the socioeconomic impacts of climate change will be. Assessing these impacts is challenged by the generally complex relationship between meteorological, biophysical and economic interactions; the expected diversity of local impacts within countries; and the relatively long-time horizon of the analysis.

The long-term local and global implications of climate change in Tunisia are projected to lead to a large total reduction in household incomes by 2030. Income reductions accumulate over time; household incomes losses are initially of US$100 million (0.4 percent of GDP), then by 2020 accumulate to US$393 million (1.4 percent of GDP) and US$1.8 billion (6.7 percent of GDP) by 2030 (see chapter 3).

Tunisia will be affected by climate change at the national and local levels, and will suffer from impacts in other countries, particularly in terms of food security. At the local level, an increase in temperatures and in some cases a reduction in precipitation is projected to reduce agricultural yields. Wheat yields for example may decrease by about 60 percent by 2050 in some parts of the Arab world. In addition, because climate change will likely reduce agricultural yields globally, world market prices for major food commodities are projected to rise. Given the high dependence of Arab countries on imported food (combined with relatively limited agricultural potential), these global dimensions are particularly important for the Arab world.

For Tunisia, there are indications that the number of droughts has increased in some regions, and will continue to become more frequent in the future. [8] Lower rainfall is likely to lead to reductions in crop yields or, in extreme cases, the complete loss of harvests, especially in rain-fed agricultural systems. Droughts also affect livestock, particularly animals that rely on pastures for feeding. It is also expected that normally occurring dry periods will last longer, exacerbating these impacts. In addition to these direct impacts on the agricultural sector and the families that rely on it, droughts also directly affect other sectors of the economy and, indirectly, nonfarm households.

Floods may also become more frequent due to climate variability and change, and as a result, induce heavy economic losses and spikes in food insecurity. Most recently, the floods in the northwestern governorates of El Kef, Bizerte, Jendouba, and Béja in early 2012 have led to a number of deaths and large economic costs. While regular flooding can be beneficial to agricultural practices in dry lands, high-magnitude flooding leads to the loss of productive land, uprooting of fruit trees, loss of livestock, and destruction of infrastructure such as irrigation facilities and rural roads. As chapter 3 clearly shows, poor rural communities are among the most vulnerable to these impacts.

Evidence suggests that the poor people in Tunisia suffer more than the nonpoor from climate change impacts, and that rural households (farm and nonfarm) have been the hardest-hit group by these adverse effects. Farmers are the most negatively affected by climate change impacts with losses of US$700 million to the economy (3 percent of 2010 GDP, see chapter 3). This high level of vulnerability of the rural poor can be explained by the joint effect of being net food buyers—who spend a high share of their income on food—and of earning incomes from climate-sensitive productive strategies, namely unskilled farm labor. It is important to note that urban households are also negatively affected by climate change in Tunisia.

Climate Change Adaptation Is about Reducing Vulnerability

Definitions and Framework

Adaptation is about reducing vulnerability. The vulnerability of countries, societies, and households to the effects of climate variability and change depends not only on the magnitude of climatic stress but also on the sensitivity and capacity of affected societies and households to cope with such stress (OECD 2009, see also box 1.4).

One conceptual framework for defining vulnerability and adaptation comes from Fay, Block, and Ebinger (2010). This framework is based on the IPCC's

Box 1.4

Definition of Climate Change Adaptation

The IPCC defines climate change adaptation as any "adjustment in natural or human systems in response to actual or expected climatic stimuli or their effects, which moderates harm or exploits beneficial opportunities." The Organisation for Economic Co-operation and Development-Development Assistance Committee (OECD-DAC) defines climate change adaptation projects as those that "reduce the vulnerability of human or natural systems to the impacts of climate change and climate-related risks, by maintaining or increasing adaptive capacity and resilience. This encompasses a range of activities from information and knowledge generation, to capacity development, planning and implementation of climate change adaptation actions." Adaptation reduces the impacts of climate stress on human and natural systems and consists of a multitude of behavioral, structural, and technological adjustments. The OECD highlights that timing (anticipatory vs. reactive, *ex ante* vs. *ex post*), scope (short term vs. long term, localized vs. regional), purposefulness (autonomous vs. planned; passive vs. active), and adapting agent (private vs. public; societies vs. natural systems) are important concepts when addressing adaptation.

Source: Based on OECD 2009 and IPCC 2001.

(2001) definition of vulnerability and seeks to capture the essence of the different concepts in the literature by defining vulnerability as a function of exposure, sensitivity, and adaptive or coping capacity (see figure 1.1).[9] As it is described in Fay, Block, and Ebinger (2010, 15): "the advantage of this approach is that it helps distinguish between what is exogenous, what is the result of past decisions, and what is amenable to policy action." This approach can be applied to communities, regions, countries, or sectors—for example, the Australian government applied this framework to agriculture.

Vulnerability is the degree to which a system is susceptible to, or unable to cope with, adverse effects of climate change, including climate variability and extremes. Vulnerability is a function of the character, magnitude and rate of climate change, and the degree to which a system is exposed, along with its sensitivity and adaptive capacity. Vulnerability increases as the magnitude of climate change exposure or sensitivity increases, and decreases as adaptive capacity increases (IPCC 2001).

The potential impact of climate variability and change on a community, sector, or system depends on exposure and sensitivity (see figure 1.1). Exposure is determined by the type, magnitude, variability, and speed of the climate event such as the changing onset of rains, minimum and maximum winter and summer temperatures, heat waves, floods, and storms. These are the exogenous factors.

Figure 1.1 Conceptual Framework for Defining Vulnerability

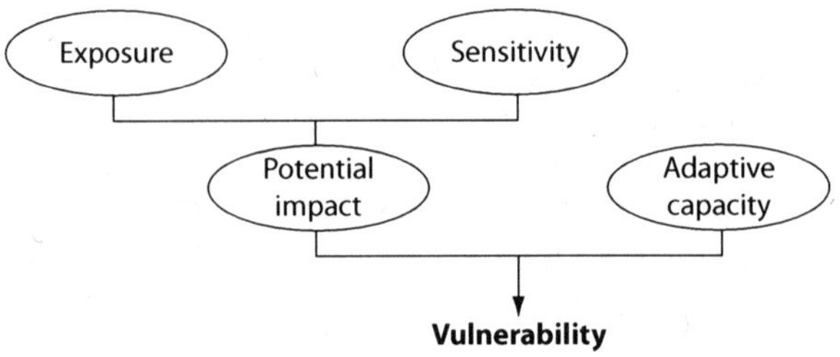

Source: IPCC 2001 (as presented in Fay, Block, and Ebinger 2010).

Sensitivity is the degree to which a system can be affected by changes in the climate, and depends in part on how stressed the system already is. Poor people and communities will be more affected than the nonpoor as they may already face stresses before a climate event. With limited assets, the poor are inherently more sensitive to even minor climate events. These are the endogenous factors.

Vulnerability depends on the severity of the potential impact and the adaptive capacity of an affected community.[10] The capacity of a system or community to adapt is determined by access to information, technology, economic resources, and other assets. It depends, moreover, on having the skills to use this information, the institutions to manage these assets, and on the equitable distribution of resources. In general, societies with relatively more equitable resource distribution will be better able to adapt than societies with less equitable distributions. This is because equitable distribution avoids resource capture, corruption and clientelism. The level of adaptive capacity tends to be positively correlated with levels of development, where adaptive capacity increases with the level of development (OECD 2009).

Climate Change Adaptation Should Be an Integrated Part of Public Sector Management for Sustainable Development

Many countries, particularly the poorest and most exposed, will need assistance in adapting to the changing climate. Urgent help is needed in preparing for drought, managing water resources, addressing sea levels, improving agricultural productivity, containing diseases, and building climate-resilient infrastructure.

How to adapt to climate change is the sovereign decision of individual countries, which includes governments, the private sector, and civil society. It is in each country's own best interest to build climate resilience and be as prepared as possible for the known and unknown consequences of climate change. This section provides a simple sketch of a framework for an adaptation process based on integrated government action. This approach, which is based on findings

Figure 1.2 Framework for Action on Climate Change Adaptation: Adaptation Pyramid

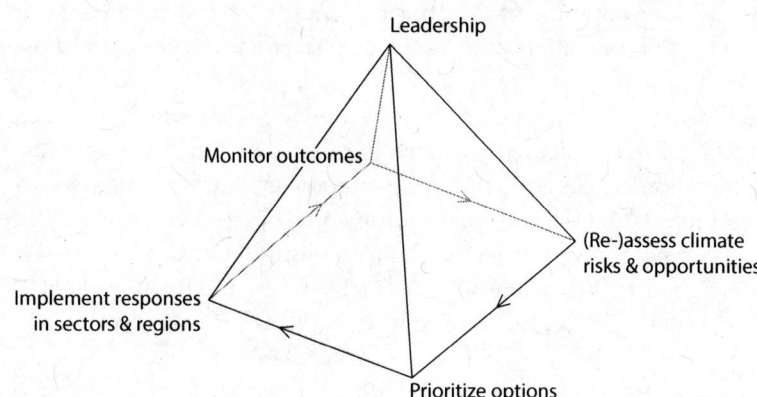

Leadership

Monitor outcomes

(Re-)assess climate
risks & opportunities

Implement responses
in sectors & regions

Prioritize options

Source: Verner 2012.

from chapters 2–4 and takes into account regional characteristics, will be developed in greater detail in chapter 5.

The prospect of climate change adds another element to be integrated into national planning. Governments, with assistance from civil society and the private sector, can ensure that a country's development policies, strategies, and action plans build resilience to a changing climate. As this report shows, an integrated approach to climate change adaptation at the country level calls for leadership, action, and collaboration and requires sound strategies to be identified, integrated, and implemented. Moreover, strategies need to be supported by legislation and action plans, including the necessary frameworks and a strong domestic policy. If they are not, these strategies can result in incoherent outcomes and maladaptation.

This report aims to provide guidance to policy makers on how to address climate change adaptation. As presented in the adaptation pyramid, there are five core components to successful adaptation (figure 1.2).

Framework for Action on Climate Change Adaptation

Adaptation is a long-term, dynamic and iterative process that will take place over decades. Decisions will need to be made despite uncertainty about how both society and climate will change. Adaptation strategies and activities will need to be revised as new information becomes available. Many standard decision-making methodologies are inappropriate, and alternative, robust methods for selecting priorities within an adaptive management framework will be more effective.

Elements in an adaptive management model include the following: (1) management objectives that are regularly revisited, and accordingly revised, (2) a model or models of the system being managed, (3) a range of management choices, (4) the monitoring and evaluation of outcomes, (5) a mechanism for incorporating learning into future decisions, and (6) a collaborative structure for stakeholder participation and learning. This model is commonly used in many fields, but particularly so in environmental policy.

In addition, and complementary to an adaptive management approach, the OECD (2009) has highlighted five enabling conditions that support the successful integration of climate change adaptation into development processes. These conditions help to ensure that multiple perspectives are brought into the policy decision process and therefore help to ensure that policy solutions that are tried are robust, and in line with an inclusive management approach. The five enabling conditions are as follows:

- A broad and sustained engagement with, and participation of, stakeholders, such as government bodies and institutions, communities, civil society, and the private sector
- A participatory approach with legitimate decision-making agents
- An awareness-raising program on climate change for households, civil society organizations, opinion leaders, and educator
- Information gathering to inform both national and local-level adaptation decisions
- Response processes to short- and long-term climatic shocks.

The adaptation pyramid (figure 1.2) provides a framework to assist stakeholders in Tunisia in integrating climate risks and opportunities into development activities. It is based on an adaptive management approach, but also highlights in particular the importance of leadership, without which adaptation efforts are unlikely to achieve what is necessary to minimize the impacts of climate change.

The base of the pyramid represents the four iterative steps that form the foundation for sound climate change adaptation decision making, namely

- Assess climate risks, impacts, and opportunities for climate action
- Prioritize policy and project options
- Implement responses in sectors and regions
- Monitor and evaluate and subsequently reassess climate risks, impacts, and opportunities.

The arrows on the four sides of the pyramid highlight the iterative nature of adaptation decision making. Adaptation is a continuous process that takes place over time and adaptation activities will be subject to revisions as new information becomes available. To this we add a fifth apex: leadership, without which adaptation efforts are unlikely to achieve what is necessary to minimize the impacts of climate change. The base of the pyramid is elaborated in the following sections.

Assess Climate Risks, Impacts, and Opportunities

In this first step, a wide range of quantitative and qualitative analyses can be used.[11] All of these rely on access to climate and socioeconomic data to provide information on climate change impacts, including on vulnerable groups, regions, and sectors. To help understand the risks and impacts, data are needed on current climate variability and change as well as projections and uncertainty about the future climate. Similarly, information on past adaptation actions and on coping strategies needs to be gathered and evaluated in light of the changing climate. Chapter 5 goes into more detail about this. Approaches to this step should be adjusted to suit the issues, locality, and circumstances.

Prioritize Options

The second step is to identify and prioritize adaptation options within the context of national, regional, and local priorities and goals and in particular the financial and capacity constraints. Expectations of climate change make it more important to consider longer-term consequences of decisions, as short-term responses may miss more efficient adaptation options or even lead to maladaptive outcomes, for example the further development of highly vulnerable locations. An effective approach is Robust Decision Making, which seeks to identify choices that provide acceptable outcomes under many feasible scenarios of the future.

Implement Responses in Sectors and Regions

Adaptive response may be somewhat at odds with immediate, local priorities, and thus the third step of implementing the agreed responses needs cooperation and understanding at national, sectoral, and regional/local levels (often jointly). At the national level, adaptation needs to be integrated into national policies, plans and programs, and financial management systems. This includes five-year plans prepared in a number of Arab countries, as well as sustainable development and poverty reduction strategies and plans; policies, regulations and legislation; investment programs, and the budget. In addition, national adaptation strategies can guide the mainstreaming of adaptation into other national policies as well as implementation at sectoral and local levels. This could involve the formation of an interministerial committee at various levels with the participation of civil society, the private sector, and academia.

Climate change needs to be considered in all sectoral activities, particularly in climate vulnerable sectors such as agriculture, health, tourism, and water. It is of paramount importance that sectoral plans and strategies take intersectoral interactions into account. For example, water is one of the key sectors for successful adaptation, but any policies concerning water management will necessarily impact agriculture and energy (irrigation), city planning (drinking water and wastewater), gender (women and girls' time working in fields and watching animals, see box 1.5), and health (waterborne diseases).

The local level is ultimately the level at which climate change impacts will be felt and responded to. Rural livelihoods tend to be anchored in climate-sensitive sectors such as agriculture, while urban livelihoods tend toward service sectors. Changing agricultural productivity due to climate change may increase food prices for those that do not produce food and even disrupt food supplies. This may also accelerate the rural to urban migration often initiated by outmigration of men from rural areas, creating challenges for rural women left behind and putting pressure on services in rapidly expanding urban areas.

Monitor Outcomes

Monitoring is essential to ensure that adaptation related strategies and activities have the intended adaptation outcomes and benefits. Comprehensive qualitative and quantitative indicators can help project proponents recognize strengths and weaknesses of various initiatives and adjust activities to best meet current and future needs. The monitoring framework should explicitly consider the effects of future climate change, particularly for projects with a long time horizon.

Adaptation Is an Iterative Process

The next step will thus be to reassess activities while taking into account new and available information, for example, about future climate change or the effectiveness of previously applied solutions.

Leadership Is Critical for Successful Adaptation

Effective climate change adaptation will not occur without strong leadership. International experience shows that the lead needs to be taken at the national level by a prominent ministry or senior government champion, such as the Prime Minister, Minister of Planning or Economy, or State Planning Commission. This champion will also require the support of a strong team comprised of representatives of relevant ministries, governorates, local authorities and institutions, the private sector, academia, civil society organizations, and ideally from opposition parties in order to ensure continuity, as government may change. Clearly, this should be adapted to the context of individual Arab countries and their specific circumstances. Leaders are needed at other levels of government and within civil society and private sector organizations. Leaders from all sectors need support through information access, education opportunities and must be treated as legitimate agents in decision-making processes. Finally, the leadership must interact with other states with regard to intergovernmental issues.

What part of the Pyramid does this report address? It provides information related to the "Assessment of Climate Change Risks and Opportunities" (chapters 2, 3, and a part of 4) and it moves toward the "Prioritization of Options" (chapters 5 and parts of 4).

Box 1.5

Gender and Climate Change in Tunisia

Climate change impacts in Tunisia are not gender neutral. Specific inequalities in men and women's access to the assets, opportunities, and decision-making power that would enable them to successfully adapt to new climate conditions and the differential social roles of men and women in Tunisia, particularly in rural areas, result in differential vulnerabilities and adaptive capacities. To strengthen resilience to climate change, it is essential to build a holistic and gender-responsive approach to adaptation that empowers women as agents of change.

The drivers of gender-based vulnerability to climate change can be separated into three general areas of inequality: access to resources, opportunity for improving existing livelihoods and developing alternative livelihoods, and participation in decision making. In the rural areas of Tunisia, it is most often women, and especially poor women, who face structural inequalities and socio-cultural norms that disadvantage them in these three areas. This intensifies their exposure and sensitivity to climate change impacts. As a result, rural women are more likely to have lower adaptive capacity than men. Their lower adaptive capacity results in exacerbated welfare impacts on individuals, households, and communities.

Significant progress has been made in recognizing the links between climate change and gender at the international level. In 2007, the United Nations and 25 international organizations formed the Global Gender and Climate Alliance (GGCA), which aims to ensure that global climate policies are gender responsive. The IPCC now recognizes gender as one factor that shapes vulnerability to climate change. In 2010, the Cancun Agreements recognized gender equality as integral to adaptation. At COP-17 in 2011, references to gender and women were strengthened in a number of important areas, for instance in countries' guidelines for National Adaptation Programmes of Action (NAPAs), in the Nairobi Work Programme,[12] and in the operationalization of the Cancun Agreements, including the Green Climate Fund, the Adaptation Committee, the Standing Committee on Finance, and the Technology Mechanism (Arend and Lowman 2011; WEDO 2011). Greater efforts can be made to incorporate gender considerations into Tunisia's climate change response.

Source: World Bank 2012, Chapter 7, "Gender-responsive Climate Change Adaptation: Ensuring Effectiveness and Sustainability in Adaptation to A Changing Climate in the Arab World."

Notes

1. People living in cities and those working in tourism are also affected by the changing climate. Climate change also affects gender dynamics and people's health. These areas are addressed for the Arab countries in Verner (2012).

2. IPCC (2007) points out that there are "sharp differences across regions and those in the weakest economic position are often the most vulnerable to climate change and are frequently the most susceptible to climate-related damages, especially when they face multiple stresses. There is increasing evidence of greater vulnerability of specific groups…" IPCC (2007) makes specific mention of traditional peoples and ways of living only in the cases of Polar Regions and small island states.

3. See for example, Verner (2012), Mearns and Andrew (2009), and Kronik and Verner (2010).

4. This report does not address in detail topics that are being addressed extensively in other ongoing work in Tunisia or the MENA region in the World Bank. For example, climate migration and biodiversity and ecosystems.

5. This report focuses on adaptation as per the request of the Ministry of Environment, as other tasks are ongoing in the energy sector in Tunisia that address mitigation.

6. *Observatoire du Niveau de la Mer.*

7. Findings are based on qualitative analyses and quantitative modeling (Computable General Equilibrium—CGE).

8. A drought is defined as an extended period (months to years) during which a region receives consistently below average precipitation leading to low river flows, reduced soil moisture and thus adverse impacts on agriculture, ecosystems, and the economy.

9. An overview of adaptation frameworks is given in Füssel 2007.

10. Sensitivity and adaptive capacity are usually inversely correlated as shown for countries in Eastern Europe and Central Asia in Fay, Block, and Ebinger (2010).

11. For an overview of available tools to assist in climate risk analysis please see http://climatechange.worldbank.org/climatechange/content/note-3-using-climate-risk-screening-tools-assess-climate-risks-development-projects.

12. The Nairobi Work Programme assists developing countries to "improve their understanding and assessment of impacts, vulnerability and adaptation to climate change" and "make informed decisions on practical adaptation actions and measures", see http://unfccc.int/adaptation/nairobi_work_programme/items/3633.php.

References

Arend, Elizabeth, and Sonia Lowman. 2011a. "Governing Climate Funds that Will Work for Women." Research Report, Women's Environmental & Development Organization, Oxfam, September 2011. http://www.genderaction.org/publications/11/climate-funds-for-women.pdf

Fay, Marianne, Rachel I. Block, and Jane Ebinger. 2010. *Adapting to a Climate Change in Eastern Europe and Central Asia.* Washington, DC: World Bank.

Füssel, H. -M. 2007. "Vulnerability: A Generally Applicable Conceptual Framework for Climate Change Research." *Global Environmental Change* 17 (2): 155–67.

IPCC (Intergovernmental Panel on Climate Change). 2001. "Climate Change 2001: Impacts, Adaptation and Vulnerability." Contribution of Working Group II to the Third Assessment Report of Intergovernmental Panel on Climate Change, Cambridge University Press, Cambridge, UK.

———. 2007. "Climate Change 2007—The Fourth Assessment Report of the Intergovernmental Panel on Climate Change." Cambridge University Press, Cambridge, UK.

Kronik, Jakob and Dorte Verner. 2010. "Indigenous Peoples and Climate Change in Latin America and the Caribbean", World Bank, Washington, DC.

Mearns, Robin, and Norton Andrew. 2009. *The Social Dimension of Climate Change: Equity and Vulnerability.* Washington, DC: World Bank.

OECD (Organisation for Economic Coordination and Development). 2009. "Policy Guidance on Integrating Climate Change Adaptation into Development Cooperation." Organisation for Economic Coordination and Development, Paris. http://www.oecd. org/document/40/0,3343,en_2649_34421_42580264_1_1_1_1,00.html.

Stern, Nicholas. 2007. *Stern Review on the Economics of Climate Change.* Cambridge, UK: Cambridge University Press.

Tolba, Mostafa K., and Najib W. Saab. 2009. "Arab Environment Climate Change— Impacts of Climate Change on Arab Countries." Arab Forum for Environment and Development, Beirut, Lebanon.

UNDP (United Nations Development Programme). 2011. "Human Development Report 2011—Sustainability and Equity: A Better Future for All." United Nations Development Programme, New York. http://hdr.undp.org/en/reports/global/ hdr2011/.

Verner, D., ed. 2012. "Adaptation to a Changing Climate in the Arab Countries: A Case for Adaptation Governance and Leadership in Building Climate Resilience." MENA Development Report, World Bank, Washington, DC.

———. Forthcoming. *Syria Rural Development in a Changing Climate.* Washington, DC: World Bank.

WEDO (Women's Environmental & Development Organization). 2011b. "The Outcomes of Durban COP 17: Turning Words into Action." Women's Environmental & Development Organization, New York, Friday, December 16, 2011. http://www. wedo.org/themes/sustainable-development-themes/climatechange/the-outcomes-of- durban-cop-17-turning-words-into-action.

World Bank. 2011. *World Development Indicators 2011.* Washington, DC: World Bank. http://hdr.undp.org/en/media/HDR_2010_EN_Table1_reprint.pdf.

CHAPTER 2

A Synthesis of Climate Change Scenarios and Impacts

Photograph by Dorte Verner

The first step in the framework for adaptation decision making introduced in chapter 1 and elaborated in chapter 5 is to assess climate change risks and opportunities. A comprehensive analysis of climate change scenarios and projected impacts is an essential first step in understanding climate change risks in Tunisia. This chapter therefore provides an overview and assessment of projected climate change impacts on Tunisia drawing from the best science available. This chapter elaborates the first stage of the climate change adaptation pyramid (figure 1.2) by looking at how climate change science informs the assessment of climate change risk in Tunisia.

Tunisia's mean annual temperatures rose by about 1.4°C in the twentieth century, with the most rapid warming taking place since the 1970s. Annual rainfall totals have declined by 5 percent per decade in the northern part of the country since the 1950s.[1] However, these trends must be seen in the context of great variability from year to year and decade to decade.[2] Trends in extreme events are more difficult to assess, given the limited data available at the time of the study, but both warm nights and heavy rainfalls have become more frequent. Sea levels have risen across the Mediterranean by an average of more than 3 millimeters each year since 1992, although records from farther back show considerable local variability.

Regional climate change scenarios suggest that warming and drying trends could continue in coming decades, potentially exacerbating current water scarcity and putting more stress on agriculture. According to climate models, mean temperatures could rise by 1.4–2.5°C and precipitation could fall by 5–15 percent by the 2050s. Local and seasonal changes in temperature and precipitation could be even greater. Sea levels could rise by between 3 and 61 centimeters this century, depending on local heat and salinity levels in the Mediterranean.

Two statistical downscaling techniques were demonstrated for the city of Tunis. Daily maximum temperatures and precipitation scenarios were downscaled from the United Kingdom's Met Office Hadley Centre Coupled Model version 3 (HadCM3) climate model for two emissions scenarios (SRES A2 and B2) for 1961–2099. Relative to an ensemble of downscaled scenarios provided by the University of Cape Town (UCT), the HadCM3 scenarios were at the lower end of warming but more extreme end of regional drying. Depending on the downscaling method and climate model used, Tunis' annual mean maximum temperature could rise by 1.5–2.6°C, and annual precipitation totals could rise 5 percent or fall 20 percent, by the 2050s.[3]

This chapter provides a synthesis of evidence of climate variability and change in Tunisia to

- Inform national water and agricultural sector planning in Tunisia by tracking past climate trends and presenting a range of future climate change scenarios
- Illustrate statistical downscaling techniques as a means of repairing meteorological data sets and assessing sensitivity to local climate change
- Summarize potential impacts and uncertainties associated with climate change that may affect Tunisia's water, land, agriculture, and coastal zone management.

In addition to changes in temperature and precipitation, rising sea levels could increase the risk of saline water entering coastal aquifers, decreasing the amount of freshwater available to Tunisia's coastal farmers and urban areas. Other studies suggest that crop yields could be affected by changes in growing seasons and amounts of water available. Increased evaporation combined with reduced rainfall could exacerbate the salinity of Tunisia's soil, while more violent precipitation events could exacerbate erosion on slopes and sedimentation of reservoirs.

Climate change could also accelerate shortages and overuse of freshwater. However, population and economic growth could put more stress than climate change on water resources in the short and medium term. Possible exceptions include situations where a tipping point such as the limit to rainfed agriculture (about 200 millimeters per year) or perennial surface drainage (about 400 millimeters per year) is being approached. Given even modest changes in mean climate combined with large year-to-year variability, these thresholds could be reached relatively quickly, with limited time available to adapt.[4]

This chapter provides information that can inform the assessment of climate risks and opportunities. (See the Framework for Action on Climate Change Adaptation—figure 1.2—in chapter 1 of this volume.) It is worth mentioning that the World Bank is preparing Country Risk and Adaptation Profiles and a Climate Change Knowledge Portal. These are resources with datasets needed for impact and risk assessments.

Finding Evidence of a Changing Climate

This chapter's synthesis of historical trends and regional climate change scenarios was compiled using only public domain information and secondary data sources:

1. **Published syntheses** produced by national ministries, meteorological agencies, and agriculture research stations, subject to quality assurance and cross-checking with primary data analysis where feasible. For example, Tunisia's First National Communication (2001) to the United Nations Framework Convention on Climate Change (UNFCCC) provides contextual information and a preliminary climate vulnerability analysis.
2. **Secondary data sources and syntheses** of regional case studies as in Zereini and Hötzl (2008). For example, Shahin (2007) provides a compendium of meteorological, hydrological, and water quality data for stations across the Arab region.
3. **Historical station records** held in global archives such as the Royal Netherlands Meteorological Institute (KNMI) Climate Explorer, and the National Oceanic and Atmospheric Administration's National Climatic Data Center (NOAA-NCDC) Global Surface Summary of Day[5] provide daily temperature and rainfall records for the region (appendix A). HadCRUT3 gives monthly mean temperature series used in calculations of global mean temperature (latest release gives data to January 2010), and the Global Historical Climatology Network Version 2-NCDC (GHCN2) holds monthly rainfall totals. The Tunisian General Direction for Water Resources of the Ministry of Agriculture (DGRE)[6] holds monthly precipitation data (figure 2.2).
4. **Merged satellite-gauge products** such as the Global Precipitation Climatology Project (GPCP). Also, the Tropical Rainfall Measuring Mission (TRMM) multisatellite precipitation analysis provides three-hourly, $0.25° \times 0.25°$ latitude-longitude resolution data in near real time since 1998 (Huffman et al. 2007).

5. **Research publications** such as Alexander et al. (2006) provide information on trends for derived indices of climate extremes for sites across North Africa. Observed climate indices are available for a few stations in Tunisia via the European Climate Assessment and Dataset[7] (see Haylock et al. 2008). Tunisian research literature yields local data for a variety of sectors including water (for example, Leduc et al. 2007; Nasr et al. 2008), land, and agriculture (for example, Gargouri et al. 2008; Jebari et al. 2010).

6. **Monthly mean gridded climate variables** such as the Climatic Research Unit (CRU) TS 3.0[8] archive provides global coverage at 0.5° × 0.5° latitude-longitude resolution for the period 1901–2006 (Mitchell and Jones 2005).

7. **Global climate model output** from the Intergovernmental Panel on Climate Change (IPCC) Fourth Assessment Report (AR4) experiments may be acquired via the Climate and Environmental Retrieving and Archiving (CERA) portal[9] or the Climate Wizard portal.[10] Downscaling predictor variables are available from the Canadian Climate Change Scenarios Network (CCCSN)[11] for a limited number of models. Reanalysis data (quasi-observational airflow and humidity indices) may be obtained from the same source for grid-boxes covering the region. All these data are at resolutions of hundreds of kilometres.

8. **Regional climate model outputs** from the EU ENSEMBLES project whose domain extends to North Africa (see van der Linden and Mitchell 2009). Tunisia is also included in the European and African domains of the ongoing Coordinated Regional Climate Downscaling Experiment (CORDEX).[12] The Climate Systems Analysis Group at the University of Cape Town (UCT) provides statistically downscaled precipitation and temperature scenarios for individual sites across Africa, including 18 meteorological stations in Tunisia.[13]

9. **Climate summaries** such as the country profiles of Mitchell, Hulme, and New (2002) provide long-term climate averages and seasonal means for climate model scenarios (in this case, used in the IPCC Third Assessment Report).

10. **Country Risk and Adaptation Profiles and the Climate Change Knowledge Portal** (World Bank) also provide potential resources and useful datasets needed for impact and risk assessments.

A Land of Contrasting Climates

Tunisia has three distinct climate zones. A Mediterranean climate predominates in the region stretching along the northeast edge of the Atlas Mountains, providing relatively cool, wet winters and warm, dry summers. Mountain rainfall in the north results from depressions originating in the eastern Mediterranean, Atlantic systems that cross Morocco and Algeria, and cyclones that develop east of the Alps and Pyrenees. Recent research using high-resolution atmospheric data suggests that the Atlas Mountains can generate so-called "explosive cyclones"—rapidly deepening low-pressure systems that can seriously affect shipping and coastal areas (Kouroutzoglou et al. 2011).

The central and eastern region of Tunisia is semi-arid but experiences violent convective storms in autumn and spring (low atmospheric moisture levels limit their formation in summer). This region also experiences the country's greatest temperature range (that is, has a continental climate regimen). The southwest and extreme southern region is arid with significant variability in rainfall from year to year due to the interplay over the Gulf of Gabès between hot-dry air masses from the Sahara and cooler air infiltrating from the north.

In general, the various regions' topography and distance from the coast influence the temperatures and amounts of precipitation they experience. However, local geographic variations produce a mosaic of micro-climates (Berndtsson 1989; Slimani, Cudennec, and Feki 2007). For instance, annual precipitation totals can exceed 1,000 millimeters in the northwest near Béja; vary between 400 and 500 millimeters in the vicinity of Tunis and Kélibia; decline to 300 and 200 millimeters in the center of the country, near Sidi Bouzid and Gabès; and fall to less than 100 millimeters near Tozeur and Remada (figure 2.1 and figure 2.2).

Average maximum temperatures exceed 40°C in July and August at the border with Libya (Ghadames) but are closer to 30°C on the northern coastline

Figure 2.1 Monthly Mean Precipitation, Maximum and Minimum Temperatures

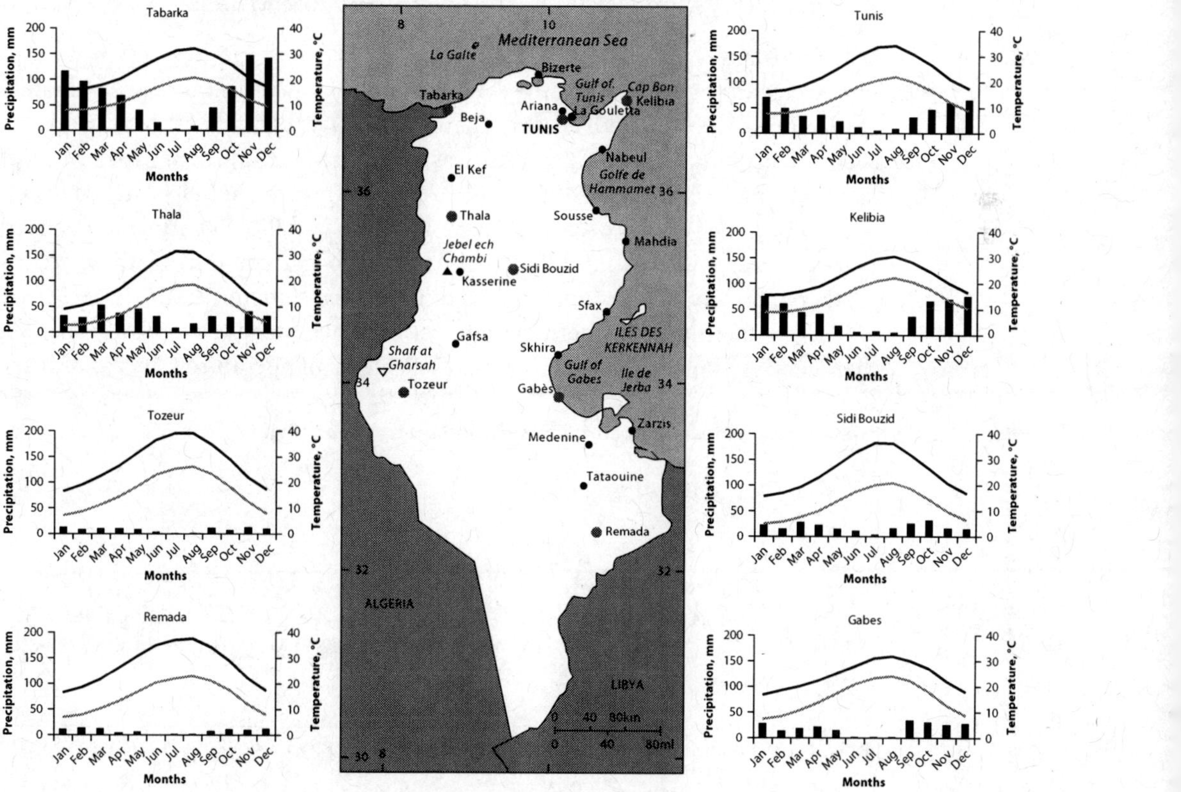

Source: Based on data from Climate Information Portal, UCT.

Figure 2.2 Mean Annual Precipitation for Northern Tunisia, 1961–2000

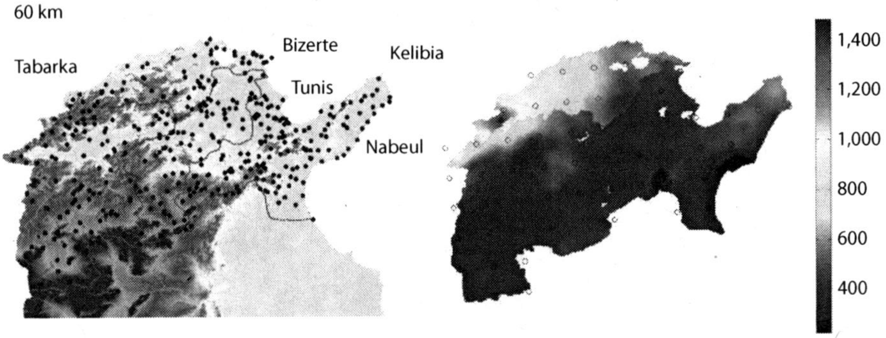

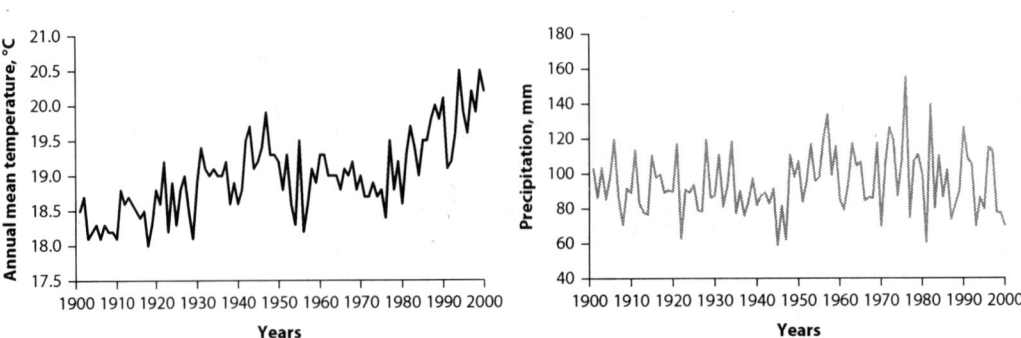

Source: Bargaoui et al. forthcoming.

(Tabarka and Kélibia) and above altitudes of 1,000 metres (Thala). Average minimum temperatures in January range from 3°C at Thala to 9°C at Kélibia on the coast. The greatest annual range is found inland where the difference between average winter and summer temperatures is more than 20°C at Tozeur.

Warmer Everywhere, Drier in the North, and Rising Seas

Temperatures

The average temperature for the whole of Tunisia increased by approximately 1.4°C during the twentieth century (figure 2.3). Overall, the most rapid warming occurred in summer (1.8°C) and the least in spring (1.2°C) (figure 2.4). Most of the warming occurred since the 1970s. However, care should be taken when interpreting such trends because of large natural variability in temperatures from year to year, as well as influence of data that do not conform to these trends (outliers), brevity, and incompleteness of individual station records (figure 2.5).

The northern and southern climate zones experienced more rapid warming than central Tunisia, but trends in annual mean temperatures since 1951 are

Figure 2.3 Twentieth-Century Mean Temperatures (Left) and Precipitation (Right)

Source: Mitchell, Hulme, and New 2002; http://www.cru.uea.ac.uk/~timm/data/index-table.html.

Figure 2.4 Twentieth-Century Seasonal Mean Temperatures

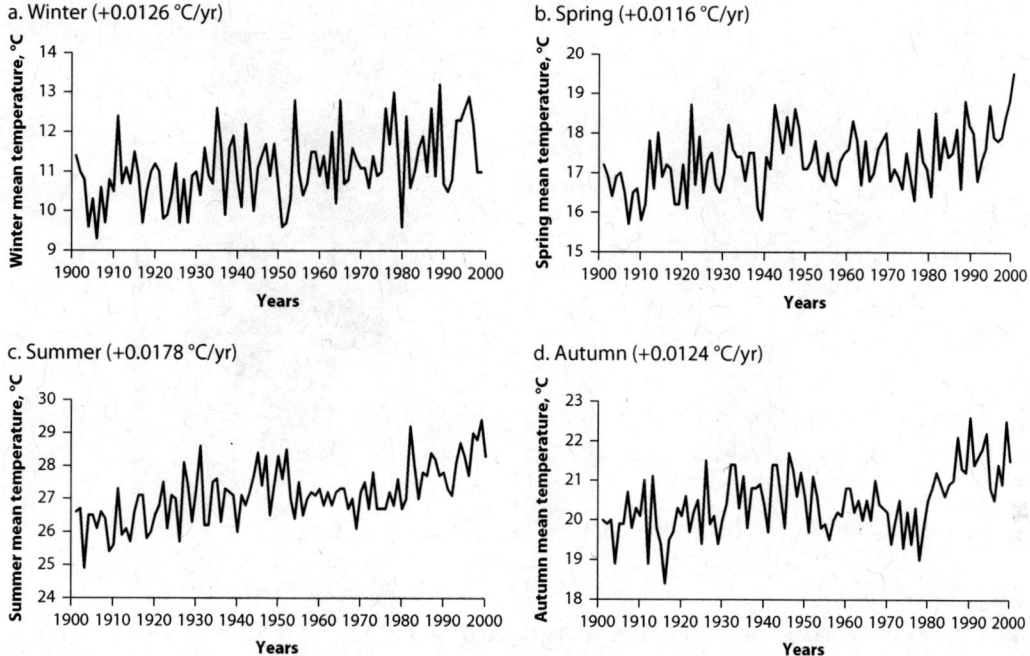

a. Winter (+0.0126 °C/yr)

b. Spring (+0.0116 °C/yr)

c. Summer (+0.0178 °C/yr)

d. Autumn (+0.0124 °C/yr)

Source: Mitchell, Hulme, and New 2002; http://www.cru.uea.ac.uk/-timm/data/index-table.html.
Note: All trends are statistically significant (*p*<.05).

Figure 2.5 Annual Mean Temperatures for Selected Stations

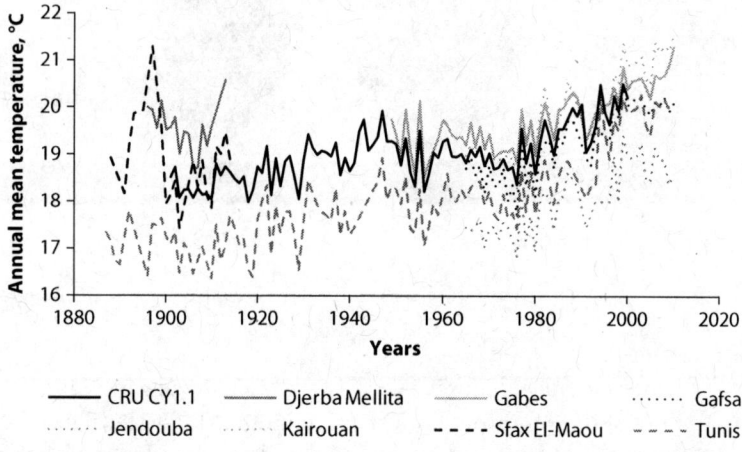

| —— CRU CY1.1 | —— Djerba Mellita | —— Gabes | ······· Gafsa |
| ······· Jendouba | ······· Kairouan | - - - - Sfax El-Maou | - - - Tunis |

Source: Based on HadCRUT3 archive.

statistically significant everywhere (figure 2.6). Overall, the greatest seasonal warming has occurred in autumn, south of a line joining Tozeur and Remada (figure 2.7). Local warming in the vicinity of Tunis was approximately 3°C during the twentieth century.

Figure 2.6 Regional Variations in the Trend in Annual Mean Temperature (Left) with Corresponding Statistical Significance Levels (Right) for the Years, 1951–2002

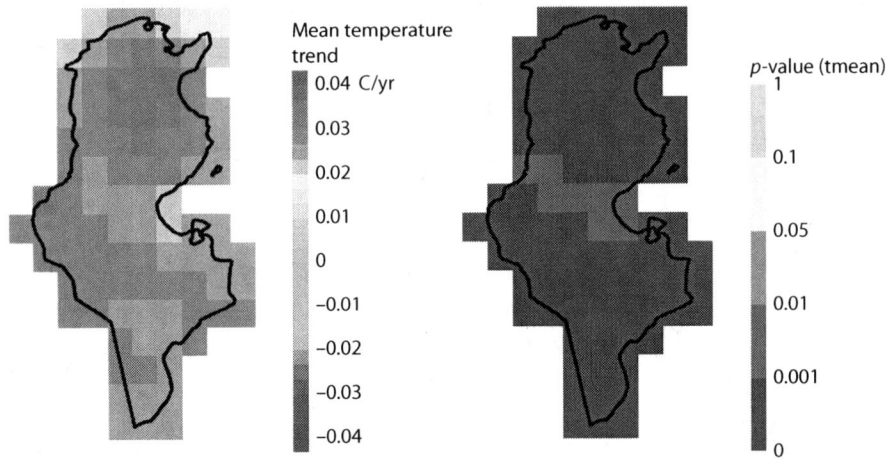

Source: Based on Climate Wizard.

Figure 2.7 Seasonal Temperature Trends (°C/yr), 1951–2002

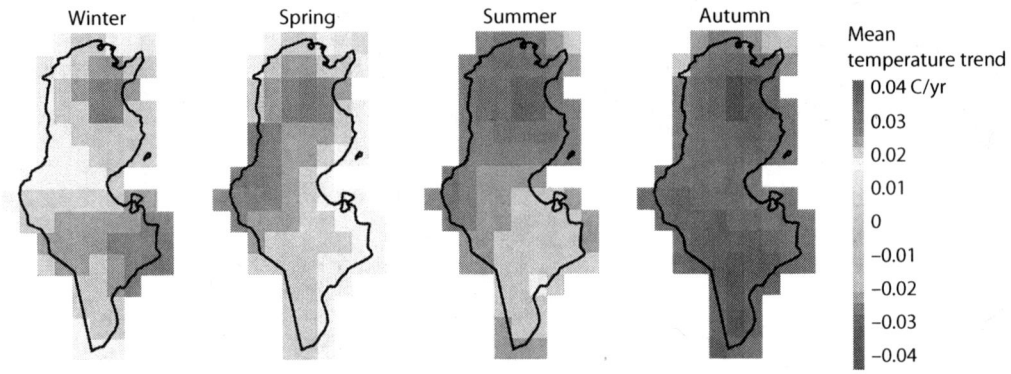

Source: Based on Climate Wizard.

Precipitation

The CRU country-average data reveal statistically significant increases in spring and summer precipitation throughout the twentieth as a whole (figure 2.8), but with great variability from year to year and decade to decade relative to trends. Therefore, such trends must be interpreted very cautiously because they depend on the period of record chosen. Unlike temperatures, more recent trends in annual precipitation totals are statistically significant only in the north of Tunisia near Tunis and Bizerte, where total annual rainfall has declined by 5 percent each decade since the 1950s (figure 2.9).

Figure 2.8 Twentieth-Century Seasonal Precipitation Totals

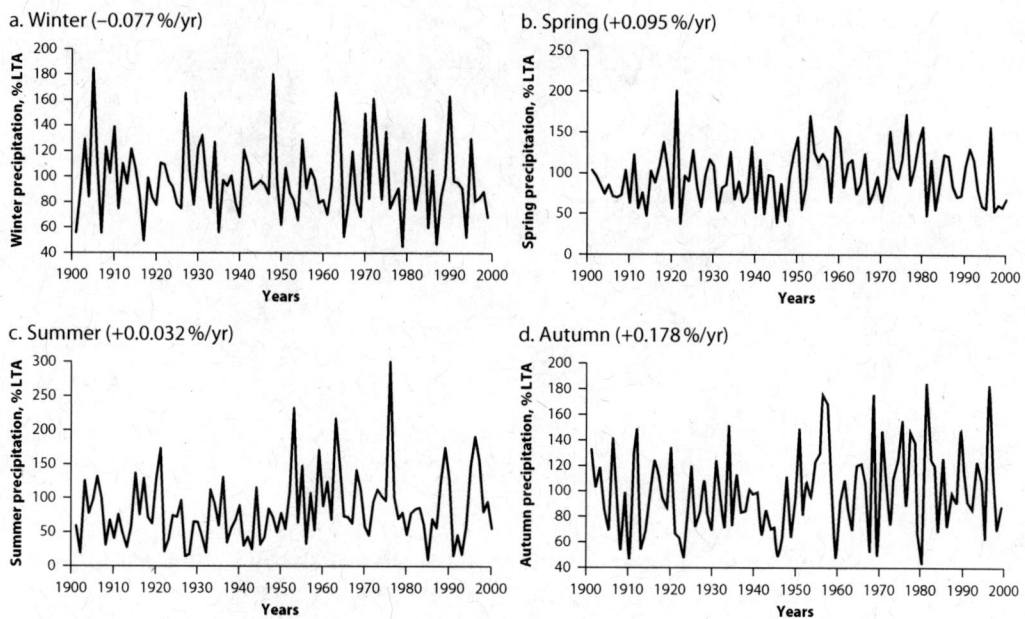

a. Winter (−0.077 %/yr)

b. Spring (+0.095 %/yr)

c. Summer (+0.0.032 %/yr)

d. Autumn (+0.178 %/yr)

Source: Based on Mitchell, Hulme, and New 2002; http://www.cru.uea.ac.uk/-timm/data/
index-table.html. Tunisia average.
Note: Trends in bold are significant at *p*<.05 LTA = long term average.

**Figure 2.9 Regional Variations in the Trend in Annual Precipitation Totals (Left) Since the
1950s with Corresponding Statistical Significance Levels (Right)**

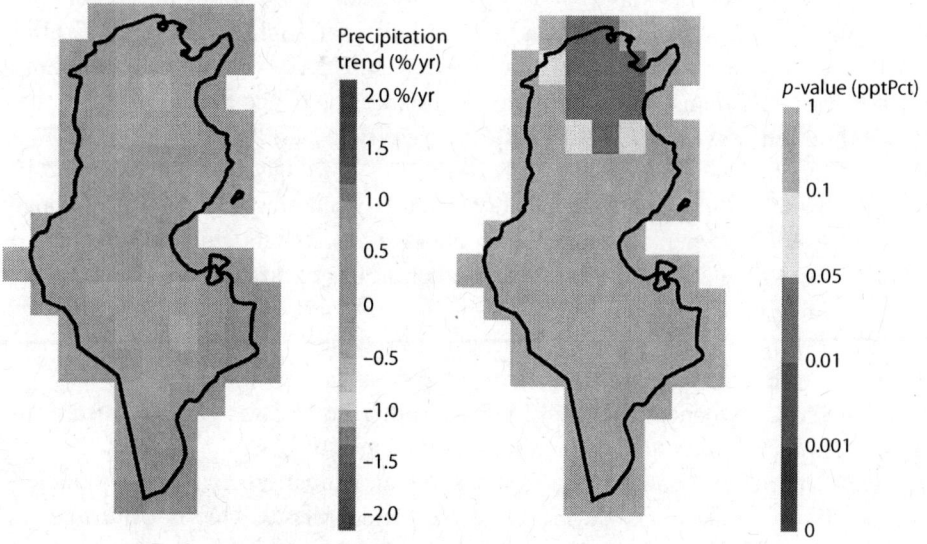

Source: Climate Wizard.
Note: pptPc t= precipitation total

Figure 2.10 Seasonal Precipitation Trends (%/yr) 1951–2002

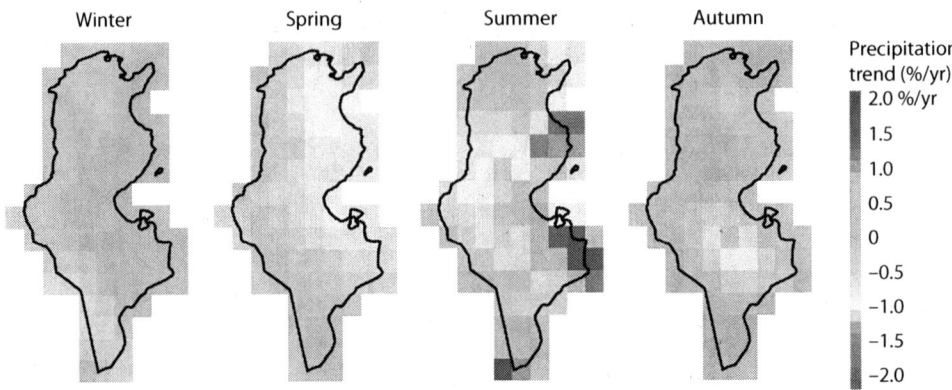

Source: Climate Wizard.

National and annual averages conceal large spatial-temporal variations in rainfall across Tunisia, especially for extreme events (see Berndtsson and Niemczynowicz 1986). For example, Kingumbi, Bargaouti, and Hubert (2005) report a notable dry period in central Tunisia during the years 1976–1989. Generally, however, western areas have experienced stable or declining rainfall, and the east has had increasing winter totals since the 1950s (figure 2.10). Conversely, spring rainfall has decreased in most areas but particularly in the eastern half of the country. Autumn rainfall has declined most in the south.

TRMM satellite data provide further evidence of variations in precipitation across Tunisia. The mean pattern for the last decade confirms the notable differences in precipitation from south to north (figure 2.10) with annual totals ranging from less than 100 millimeters on the margins of the Sahara to more than 700 millimeters on the Mediterranean coast (figure 2.11).

This pattern is accentuated in winter with a strongly negative (figure 2.12, right panel) North Atlantic oscillation (NAO), in contrast to years with strongly positive NAO, when dominant storm tracks are located much farther north over Europe (figure 2.12, left panel). Under conditions of positive NAO the pattern of winter rainfall across Tunisia is much more variable than during negative NAO winters. The influence of NAO on western North Africa has been reported before (for example, Sâadaoui and Sakka 2007).

There is evidence that the El Niño Southern oscillation (ENSO) also influences precipitation totals across Tunisia from year to year and decade to decade. For example, Dai and Wigley (2000) show an annual precipitation anomaly of up to –50 millimeters during typical El Niño phases. This is confirmed by Ouachani, Bargaoui, and Ouarda (2011), who demonstrate significant correlations between extreme ENSO and periods of severe water shortages/drought in

Figure 2.11 Rainfall Totals (Millimeters) Estimated by the TRMM System for the Decade 2001–2010 (Left Panel) and Monthly Series 1998–2011 (Right Panel)

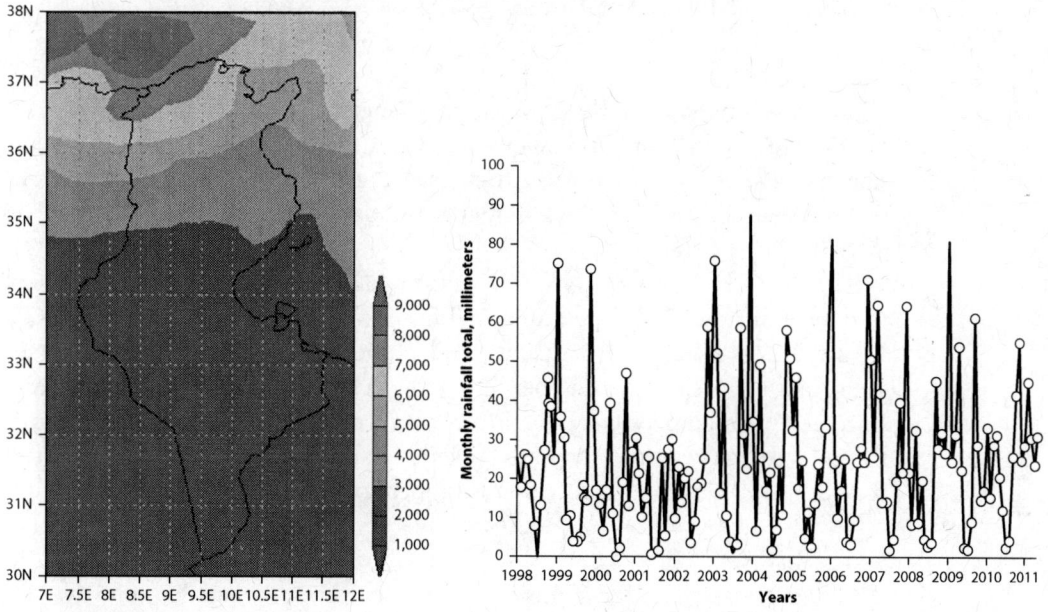

Source: TRMM Satellite Data.

Figure 2.12 Winter (December-January-February) Rainfall Totals (Millimeters) in 2007/08 (Left Panel, Positive NAO) and 2009/10 (Right Panel, Negative NAO)

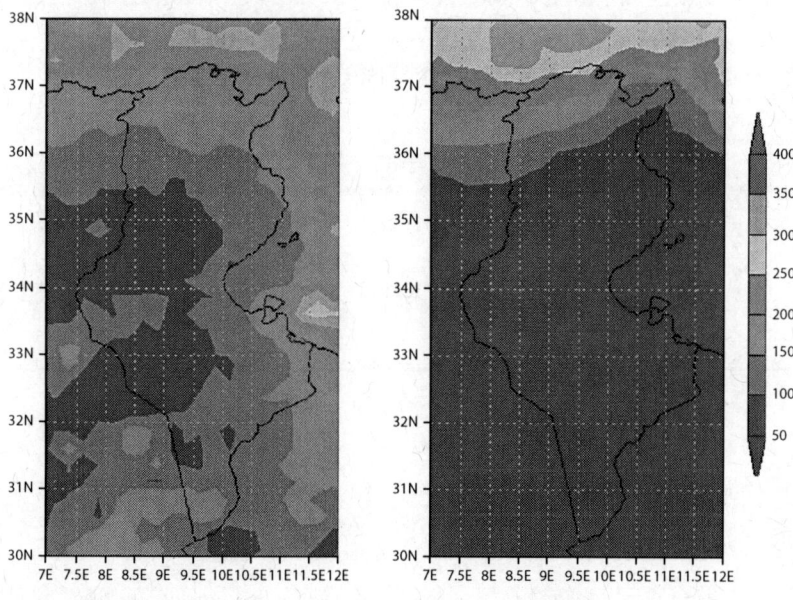

Source: TRMM Satellite Data.

northern Tunisia's upper Medjerda River basin. Likewise, Kingumbi, Bargaouti, and Hubert (2005) note an association between ENSO and negative anomalies in annual rainfall for central Tunisia.

Extreme Events

Global assessments provide evidence of changes in daily climate extremes of temperature and precipitation. For example, Alexander et al. (2006) show that since the 1950s, western North Africa has experienced significantly fewer cold nights, while the number of warm nights and days with heavy precipitation has increased.

These observations are consistent with available records. For example, summer mean maximum temperatures in Tunis have risen steadily since the 1960s (figure 2.24). Kingumbi, Bargaouti, and Hubert (2005) report a statistically significant increase in the number of days with heavy rainfall (>30 millimeters) since 1989, based on eight rainfall records in central Tunisia. The Dartmouth Flood Observatory has documented several fatal flash floods since 1986 but the record is too short to infer any trends (appendix B).

Sea Level Rise

Although tidal data for Tunisia are not currently available publicly, the University of Colorado provides monthly data on the mean sea level for the whole Mediterranean basin (figure 2.13). This index shows that the mean sea level across the basin has been rising by roughly 3.1 millimeters per year since 1992—the same as the estimated global average rate of sea-level rise of 3.1 (2.4–3.8) millimeters per year since 1993. Estimates based on TOPEX-Poseidon satellite altimetry and tide gauge data for 1993–2002 suggest a

Figure 2.13 Mean Sea Level Change in the Mediterranean, 1992–2011

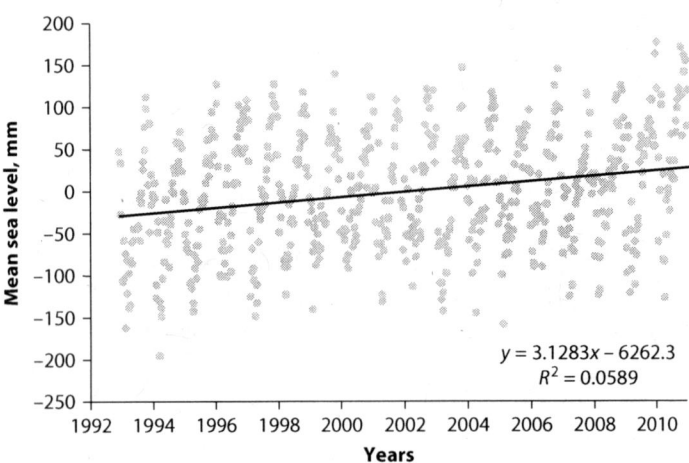

$$y = 3.1283x - 6262.3$$
$$R^2 = 0.0589$$

Source: University of Colorado Sea Level Research Group; http://sealevel.colorado.edu/content/regional-sea-level-time-series.

Mediterranean-wide sea level rise in coastal areas of 4.54 ± 0.30 millimeters per year, compared with 4.28 ± 0.30 millimeters per year in the open ocean (Mangiarotti 2007).

Reconstructed sea level trends for the western Mediterranean suggest an increase of 0.5–1.0 millimeter per year from1945 to 2000 (Calafat and Gomis 2009). This demonstrates how much sea levels can vary from one decade to the next (and provides a note of caution against relying too much on short-term records) (figure 2.14). Variations in atmospheric pressure since the 1950s are thought to have depressed sea levels; without these variations, the residual trend in the basin rises to 1.2 ± 0.2 millimeters per year, according to Calafat, Marcos, and Gomis (2010), or 0.9 ± 0.4 millimeters per year according to Marcos and Tsimplis (2007). One array of tide gauges indicates that since 1990, Mediterranean sea levels have risen at a rate 5–10 times faster than the twentieth-century mean rate (Klein and Lichter 2009).

In addition to tide-level data, anecdotal and ecological evidence point to rising sea levels. For example, salt-tolerant vegetation is reportedly migrating inland along canals and drains near the Ghar El Melh lagoon (Ayache et al. (2009) and onto coastal lowlands around the Gulf of Gabes (Oueslatia 1992). Coastal land areas are also undergoing salinization, and some archaeological sites are being inundated by the sea (Oueslatia 1995). Meanwhile, the western Mediterranean became warmer and more saline during the second half of the twentieth century (Mariotti 2010; Vargas-Yanez et al. 2008, 2010a, 2010b).

Figure 2.14 MSL Reconstructed for the Mediterranean for the Period, 1945–2000

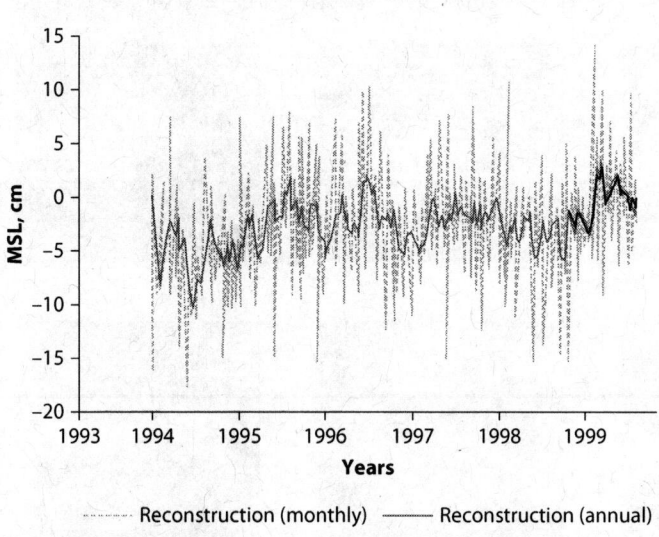

Reconstruction (monthly) ⎯⎯ Reconstruction (annual)
⎯⎯ Altimetry

Source: Reprinted from Global and Planetary Change, 66, F. M. Calafat, D. Gomis, Reconstruction of Mediterranean sea level fields for the period 1945–2000, pp 225–234, 2009, Figure 11, with permission from Elsevier.
Note: The thin gray line corresponds to the original monthly data and the thick black line is a one-year moving average. The red line corresponds to MSL computed from altimeter data for the period 1993–2000 MSL = mean sea level..

Climate Models Suggest More Local Warming and Drying

Mediterranean Basin

Tunisia lies within the Mediterranean (SEM) and North Africa (NAF) subregions of Southern Europe identified in IPCC AR4. This section focuses on the SEM subregion because more climate model results are available here. Further, the expected warming and drying of the Mediterranean basin (figures 2.15 and 2.16) have been widely discussed in the research literature (for example, Gao and Giorgi 2008; Giorgi and Bi 2005; Hertig and Jacobeit 2008; MedCLIVAR 2004).

Several studies consider the causes of a projected decline in precipitation for the region. It is speculated that global climate change could cause northern hemisphere storm tracks to move toward the North Pole and hence weaken the

Figure 2.15 Temperature and Precipitation Changes over Europe and the Mediterranean from the MMD-A1B Simulations

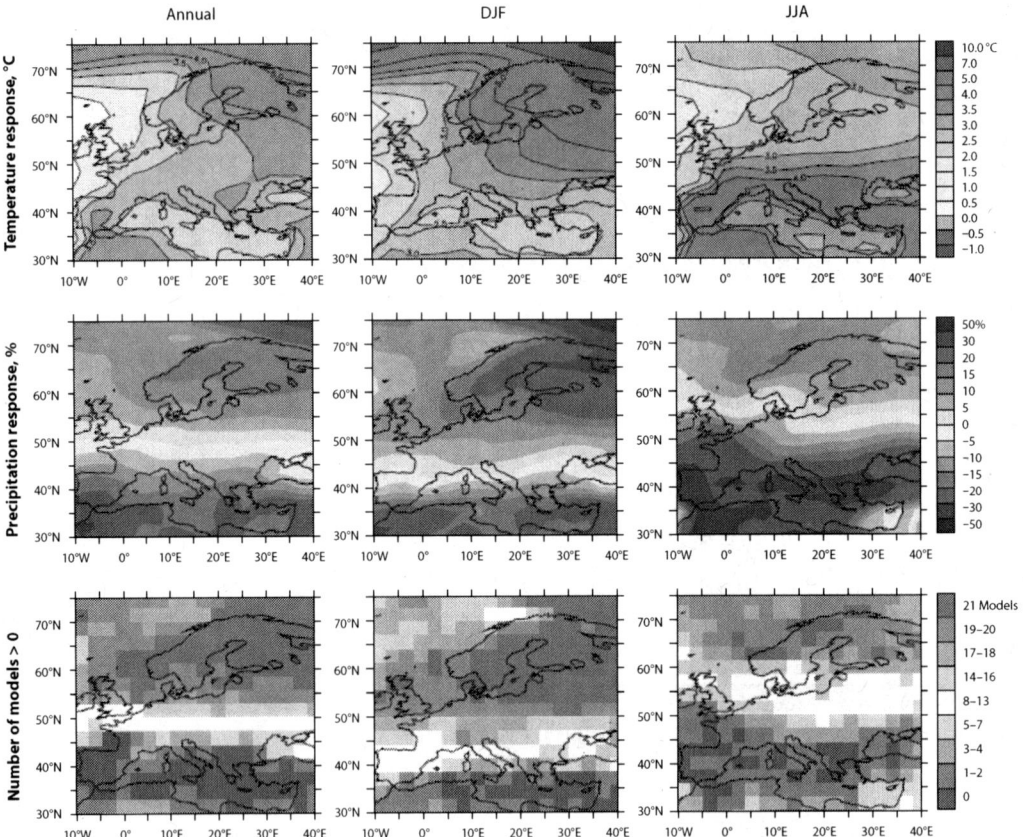

Source: Christensen et al. 2007. Top row: Annual mean, winter (DJF) and summer (JJA) temperature change between 1980–99 and 2080–99 averaged over 21 models. Middle row: same as top, but for fractional change in precipitation. Bottom row: number of models out of 21 that show precipitation increases.

Figure 2.16 Climate Change Signal (2021–50 Minus 1961–90) for Annual Near-Surface Temperature (°C) (Left Panel) and Annual Precipitation Total (%) (Right Panel) for the 16-Regional Climate Model (RCM) Mean of ENSEMBLES under SRES A1B Emissions

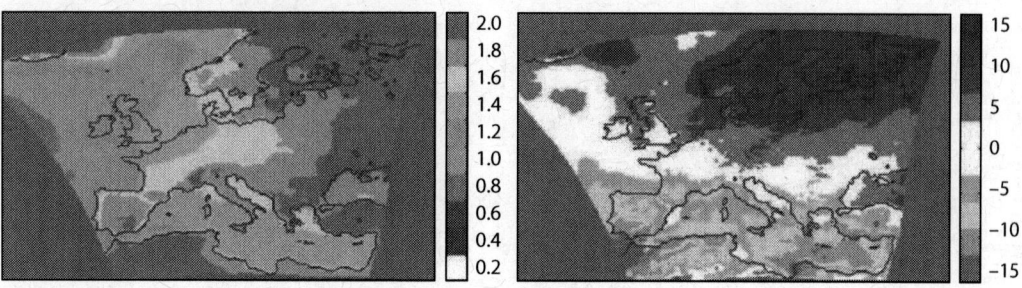

Source: van der Linden and Mitchell 2009. The ENSEMBLES work reproduced here is Figure 6.3 and Figure 6.5 from the EU-funded FP6 Integrated Project ENSEMBLES (Contract number 505539).

strength of the Mediterranean storm track (Bengtsson, Hodges, and Roeckner 2006). This is expected to reduce the number of cyclones crossing the Mediterranean (Giorgi and Lionello 2008; Lionello and Giorgi 2007). One climate model experiment projects 10 percent fewer cyclones in the western Mediterranean (Raible et al. 2010).

As previously noted, the region exhibits considerable natural climate variability from year to year and strong climatic gradients due to topographical, coastal, and marine influences. All make simulating the regional climate problematic. Although models reproduce the changes in regional temperature observed over the second half of the twentieth century, precipitation processes are not so well resolved by the present generation of climate models. For example, even under present conditions, at a model resolution of tens of kilometres, regional climate models such as RegCM3 tend to overestimate rainfall totals in mountainous areas (Pal et al. 2007). Other models within the EC's ENSEMBLES climate change project, such as DMI-ARPEGE, are known to consistently underestimate observed precipitation across northern Tunisia (figure 2.17).

Climate models project that sea level could rise in the Mediterranean by 3–61 centimeters during the twenty-first century as its waters warm and become more saline (Marcos and Tsimplis 2008). However, the combined effects of temperature and salinity are very uncertain and offer no consistent patterns. Variations in salinity are partly controlled by exchanges of water through the Straits of Gibraltar; however, this phenomenon cannot be addressed in the present generation of global climate models. Changes in atmospheric pressure are expected to have a relatively modest effect, reducing sea level by less than 0.6 centimeters, whereas ocean circulation changes could add 6 centimeters in certain areas (Tsimplis, Marcos, and Somot 2008).

The Web-based portal Climate Wizard provides data to map changes in temperature and precipitation projected by 16 general circulation models (GCMs) and three greenhouse gas emissions scenarios (A2, A1B, B1) for Tunisia by the 2050s (figures 2.18 to 2.20). Maps were produced for the climate model group

Figure 2.17 Observed and RCM Monthly Mean Precipitation Totals in North Tunisia (Cap-Bon, Medjerda Basin Upstream [U] and Downstream [D] of the Sidi Salem Dam, Northern Coastal Basins)

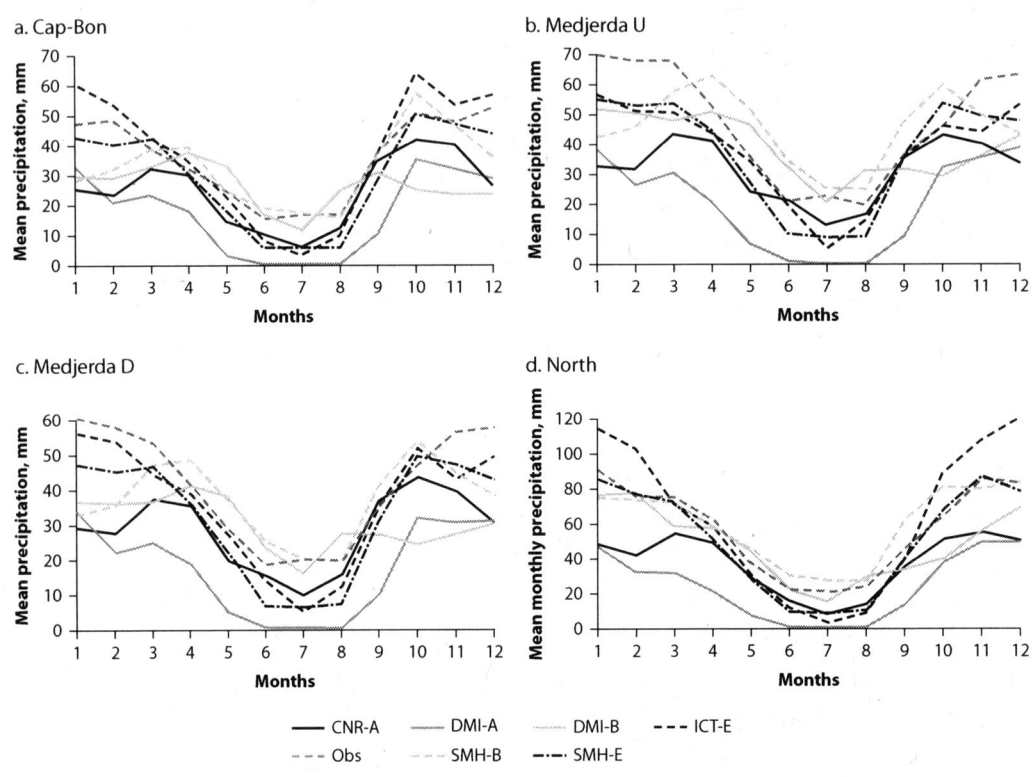

Source: Bargaoui et al. forthcoming.

(ensemble) annual and seasonal means, as well as for the most extreme models in the ensemble to demonstrate the ranges of uncertainty. Mean annual temperatures are projected to increase by 1.4–2.5°C in the 2050s (figure 2.20, upper row). Precipitation is projected to decrease by 5–15 percent over the same period (figure 2.20, lower row). These ranges are expected to be 2.4–4.2°C and 5–25 percent, respectively by the 2080s.

Seasonal mean temperatures show much greater variation when comparing the most extreme models in the ensemble. Under the SRES B1 emissions scenario, the greatest warming is expected in western and southern Tunisia in summer (figure 2.19, upper row). Under the SRES A2 emissions scenario a summer "hot spot" emerges on the border with Algeria near Tozeur (figure 2.18, lower row). Here, the local temperature increase in summer could be as great as 5.3°C by the 2050s.

Future changes to seasonal precipitation remain somewhat uncertain despite a general consensus of drying for Tunisia (see figures 2.15 and 2.16). In fact, the ensemble range brackets reductions and increases in precipitation in all seasons

Figure 2.18 IPCC AR4 Ensemble Mean Annual Changes in Temperature (Upper Row) and Precipitation (Lower Row) under Scenario B1 and A2 Emissions by the 2050s and 2080s

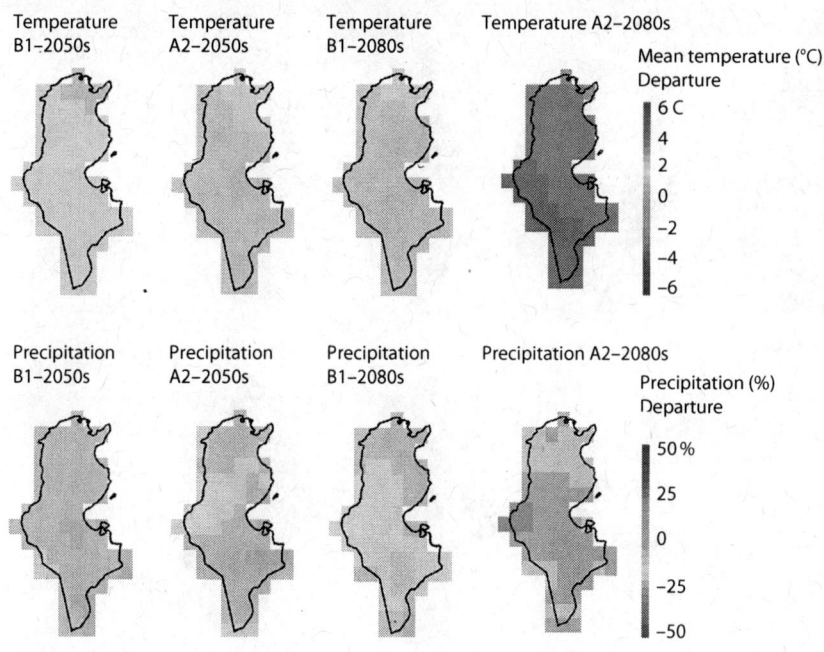

Source: Climate Wizard.

Figure 2.19 Ensemble Range (B1-Lowest, Upper Row; A2-Highest, Lower Row) for Seasonal Temperature Changes by the 2050s

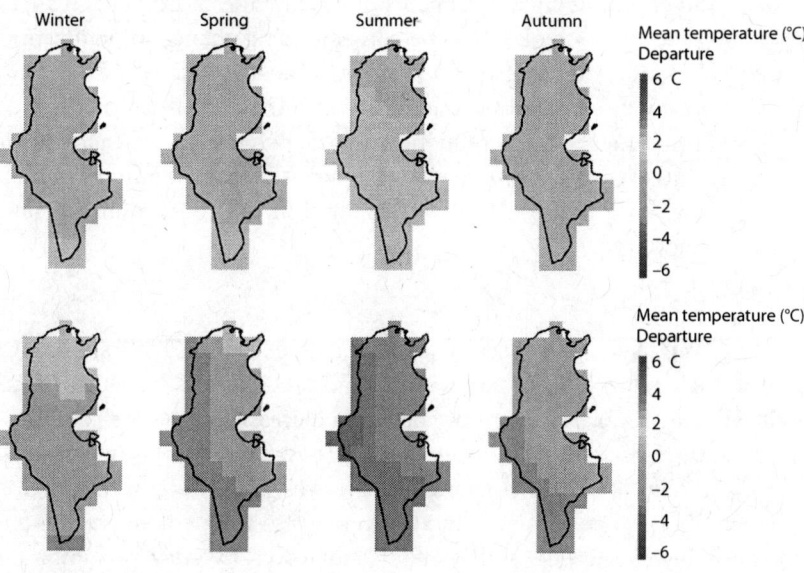

Source: Climate Wizard.

Figure 2.20 Ensemble Range (B1-Lowest, Upper Row; A2-Highest, Lower Row) for Seasonal Precipitation Changes by the 2050s

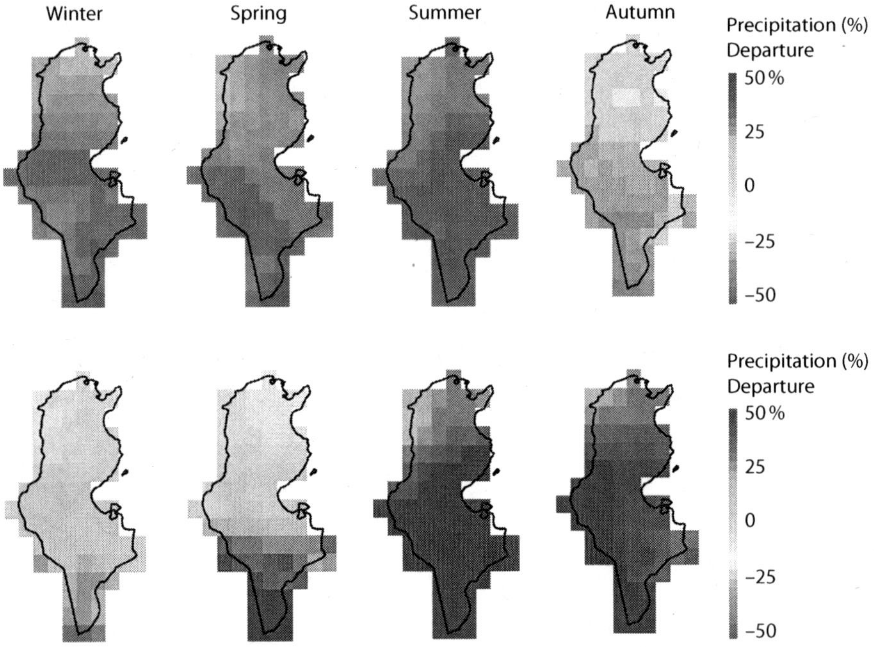

Source: Climate Wizard.

(figure 2.20). The greatest uncertainty is in summer rainfall over southern Tunisia where the ensemble figures range from a reduction of 80 percent to an increase of 250 percent by the 2050s. The corresponding change in winter precipitation totals spans a decrease of 55 percent to an increase of 35 percent. These uncertainty ranges expand to a decrease of 95 percent to an increase of 300 percent (in summer) and a reduction of 65 percent to an increase of 80 percent (in winter) by the 2080s. However, given that the amount of rainfall is already very low, (see figure 2.1 above) changes in the south translate into small absolute amounts.

Extreme Events

There has been relatively little research on changing extreme events specific to North Africa. A global analysis of temperature and precipitation extremes based on nine GCMs shows more frequent dry days and increasingly intense precipitation, but the only statistically significant changes (based on at least five models) are fewer frost days, increasingly long heat waves and more warm nights (Tebaldi et al. 2006). The same study also found less frequent heavy rainfall (daily totals of fewer than 10 millimeters) and lower five-day precipitation totals, although these were not significant according to the assessment criteria.

High-resolution simulations with RegCM3 indicate that the region could be more arid by the end of the twenty-first century and that drylands could extend north into Tunisia (Gao and Giorgi 2008). Downscaling (using different statistical techniques, configurations of predictor sets, host global climate models, and emissions scenarios) also points to decreasing precipitation throughout Tunisia (Hertig and Jacobeit 2008).

Cautionary Remarks

The expected signs of climate change in the Mediterranean basin (including Tunisia) are consistent across a range of models/scenarios (Giorgi and Lionello 2008). However, the models' consensus on this point does not mean that they are accurate predictions, because they explore regional climate responses only within a limited set of conditions (Pielke and Wilby 2012). They do not, for instance, incorporate the effects of land surface and aerosol feedbacks on the solar energy balance. Further, because the ensembles comprise models that are closely related, their results are not statistically independent. Nonetheless, global and regional climate scenarios showing fewer cyclones over the Mediterranean can be subjected to stringent testing against observed data. These observations are especially pertinent to regional climate downscaling (in the next section), which depends on the quality of the global model information for the production of local scenarios.

Building Climate Change Scenarios for Tunis, Downscaling Methods

Two statistical downscaling methods were used to construct illustrative scenarios of daily maximum temperatures (TMAX) and precipitation amounts (PRCP) for the city of Tunis. The Statistical DownScaling Model (SDSM) v4.2 (Wilby, Dawson, and Barrow 2002) is based on empirical relationships between local variables, in this case PRCP and TMAX, and large-scale atmospheric predictors such as sea level pressure, southerly wind strength, and surface relative humidity. In common with other statistical downscaling methods, these relationships are assumed to be valid under future greenhouse gas conditions. The model incorporated the observed daily TMAX and PRCP in Tunis from 1981 to 2000, then was run for the years 1961–2099 under SRES A2 and B2 emissions scenarios. Predictors centered on the grid-box to the southwest of the city had to be used because the underlying surface of the grid-box containing Tunis is ocean in the climate model (figure 2.21).

The UCT portal provided statistically downscaled scenarios for each month. The UCT system has been extensively tested using sites across Africa and against other downscaling techniques (see Hewitson and Crane 2006; Hewitson and Wilby 2008). The UCT model has already been calibrated, and preprepared scenarios may be obtained for seven global climate models using SRES A2 and B1 emissions scenarios for the 2050s (2046–65) and 2090s (2081–2100).

Figure 2.21 Land-Sea Mask of HadCM3 over the Mediterranean

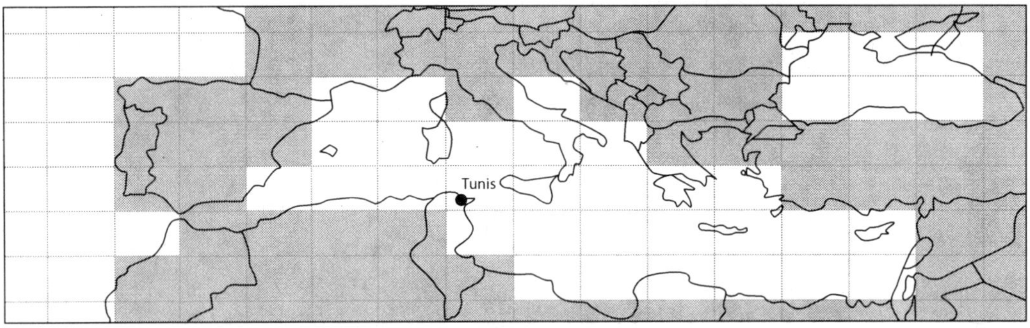

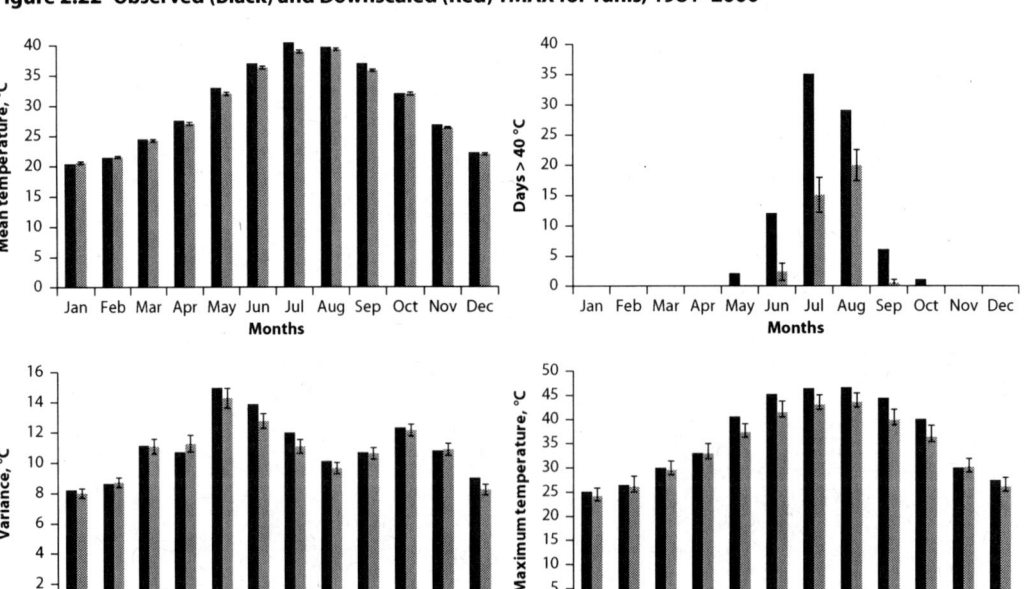

Source: Reprinted from Global and Planetary Change, Vol. 68, Issue 3, C. Giannakopoulos, P. Le Sager, M. Bindi, M. Moriondo, E. Kostopoulou, C. M. Goodess, Climatic changes and associated impacts in the Mediterranean resulting from a 2°C global warming, pp. 209–224, 2012, with permission from Elsevier.

Temperature

Figure 2.22 shows that SDSM faithfully reproduces the observed monthly mean and variance of TMAX in Tunis (left column). However, the model underestimates the frequency of daily TMAX exceeding 40°C and the absolute maximum recorded in each month (right column).

Figure 2.22 Observed (Black) and Downscaled (Red) TMAX for Tunis, 1981–2000

Source: Based on SDSM.

Figure 2.23 Observed (Black) and Downscaled (Gray Lines) Daily TMAX for Tunis in 1993 (Left Panel) and 1965 (Right Panel)

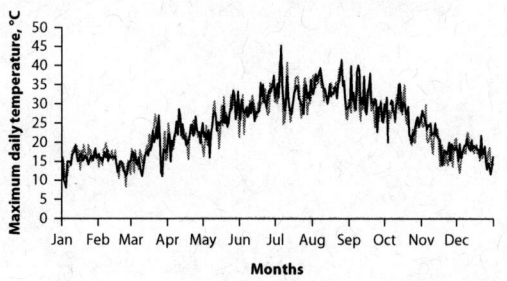 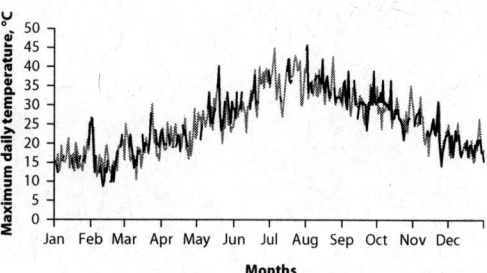

Source: SDSM Downscaled data are for an ensemble of 20 simulations.

Aside from the most extreme values, SDSM produces credible simulations of Tunis' daily and annual temperatures (figure 2.23, left). This supports the view that the model could be used to repair or fill in missing data for earlier periods (figure 2.23, right) for which there are no temperature observations, or to flag outliers within quality assurance processes.

For the purposes of climate change applications it is also necessary to assess downscaling skill over timescales of years and decades. Figure 2.24 compares observed and downscaled summer mean TMAX. Clearly, SDSM simulates year-to-year variability as well as the underlying trend of rising maximum temperatures. This confirms that the chosen downscaling variables have predictive skill over daily and decade-to-decade timescales (see also: Corte-Real, Zhang, and Wang 1995).

Illustrative TMAX scenarios were then downscaled for Tunis from the climate model HadCM3 under SRES A2 and B2 emissions scenarios for the 2020s, 2050s and 2080s. In this analysis, temperature changes were greatest in autumn (September to November), consistent with trends observed since the 1950s (figure 2.7). SDSM suggests that TMAX could rise by 2.5–3.3°C by the 2080s depending on emissions (figure 2.25). These figures lie at the lower end of the TMAX changes found in the UCT ensemble (figure 2.26).

Precipitation

Daily precipitation occurrence and amounts are more challenging to downscale than temperatures, partly because of the quality and completeness of observations, and partly because local rainfall is difficult to predict (particularly in arid and semi-arid environments) (Wilby 2008). Nonetheless, SDSM reproduces the wet season and summer minimum rainfall for Tunis (figure 2.27, top left). Observed rainfall totals also lie within the ensemble ranges for each month (figure 2.27, top right). Heavy daily rainfalls (defined here as greater than or equal to the 95th percentile) are generally underestimated (figure 2.28) but their seasonal occurrence and August peak is reflected (figure 2.27, lower left).

Figure 2.24 Observed (Black Line) and Downscaled (Gray Line) Summer Mean Maximum Temperatures (TMAX) in Tunis, 1961–2000

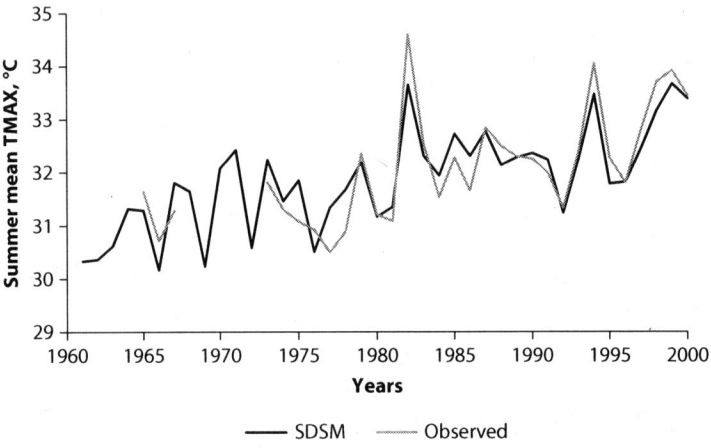

Source: Based on SDSM.

Figure 2.25 Changes in TMAX (°C) Downscaled from HadCM3 to Tunis under SRES A2 (Left Panel) and B2 (Right Panel) Emissions for the 2020s, 2050s, and 2080s

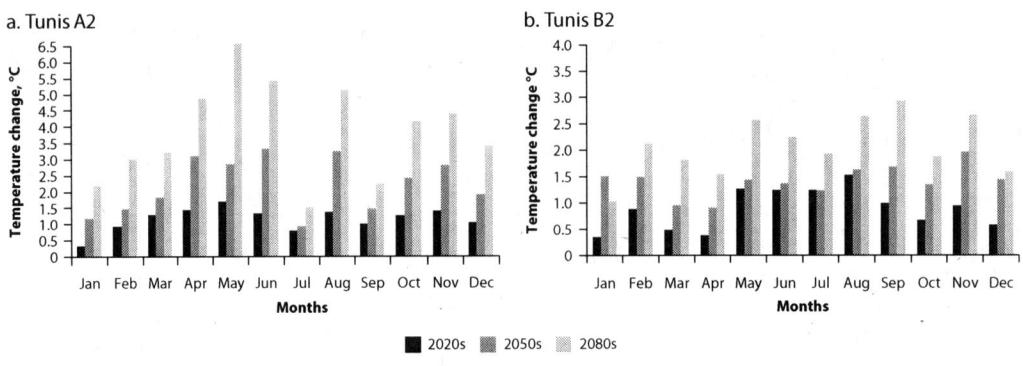

Source: Based on SDSM.

Finally, the observed average length of each month's dry spell is mirrored by SDSM (figure 2.27, lower right).

Observed and downscaled precipitation statistics would not be expected to match exactly because of the large amount of missing data (approximately 12 percent of the calibration set for Tunis). In addition, mean statistics do not adequately represent semi-arid and arid climates because of the influence of highly localized, extreme precipitation events (Berndtsson and Niemczynowicz 1986; Wilby 2008). However, figures 2.27 and 2.28 suggest that SDSM does capture key features of the precipitation given information about large-scale atmospheric conditions over the region. On this basis, illustrative precipitation

Figure 2.26 Range of Changes in Monthly Mean TMAX (°C) in Tunis

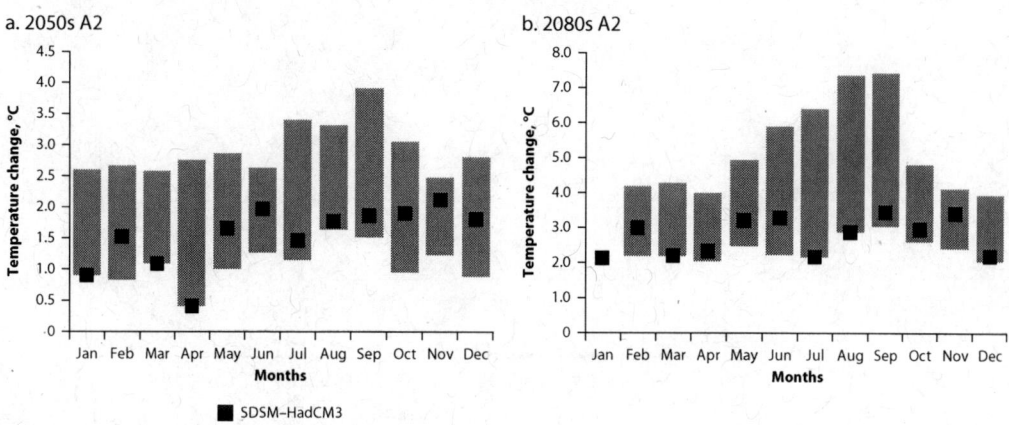

a. 2050s A2

b. 2080s A2

■ SDSM–HadCM3

Source: SDSM-HadCM3.
Note: Range of changes in monthly mean TMAX (°C) at Tunis downscaled by the UCT ensemble (pink bars) and SDSM (brown box) under SRES A2 emissions for the 2050s (left) and 2080s (right).

Figure 2.27 Observed versus Downscaled Monthly Mean Precipitation Total Based on Daily Wet-Day Probability, 1981–2000

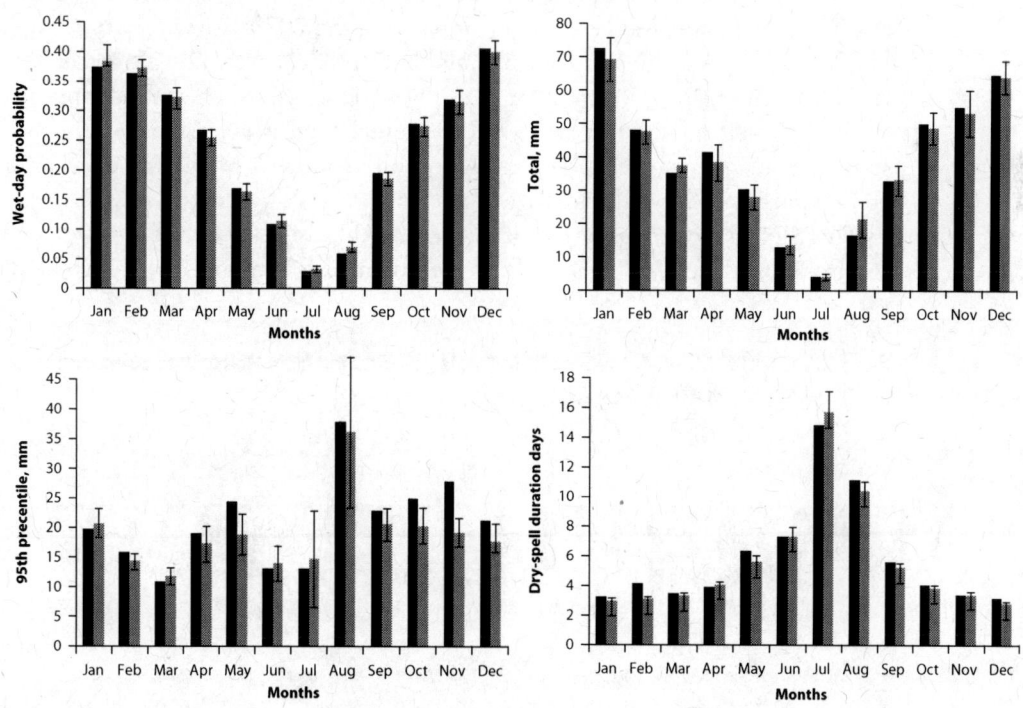

Source: Based on SDSM.
Note: Observed (black bars) and downscaled (gray bars) daily wet-day probability, monthly mean precipitation total, 95th percentile wet-day amount, and average dry-spell length in Tunis 1981–2000. T-bars show standard deviations of the downscaled ensemble.

Figure 2.28 Observed and Downscaled Distributions of Daily Rainfall Amounts in Tunis, 1981–2000

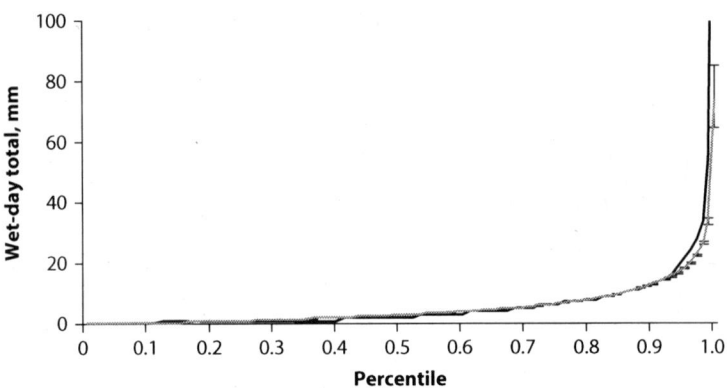

Source: Based on SDSM.
Note: Observed (black line) and downscaled (red line) distributions of daily rainfall amounts in Tunis for the period 1981–2000. T-bars show standard errors of the SDSM.

scenarios were constructed for Tunis using predictors from HadCM3 under the SRES A2 and B2 emissions scenarios.

Precipitation totals decline under both scenarios in all months and time periods. The small increases seen in a few months in the 2020s are within the range of natural variability. Overall, annual rainfall totals decline by 29–32 percent in the 2050s, and by 38–52 percent in the 2080s (figure 2.29). The near disappearance of rainfall in summer and reduced annual total is consistent with other studies showing a northward migration of the southern Mediterranean's arid zone (Giorgi and Lionello 2008)

The SDSM precipitation scenarios fall at the dry end of the UCT range (figure 3.30). The UCT ensemble also contains a few models with increased

Figure 2.29 Changes (Percent) in Monthly Precipitation Totals Downscaled from HadCM3 to Tunis under SRES A2 (Left Panel) and B2 (Right Panel) Emissions

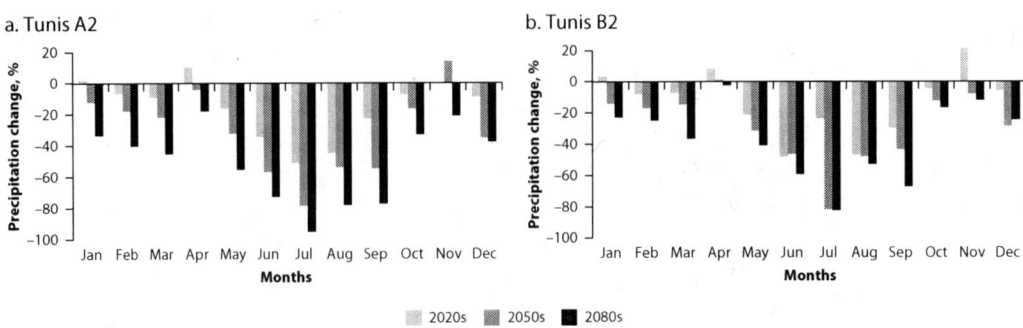

Source: Based on SDMS.

Figure 2.30 Percentage Changes (Percent) in Monthly Precipitation Totals in Tunis Downscaled from the UCT Ensemble (Light Bars) and SDSM (Dark Box) under SRES A2 Emissions

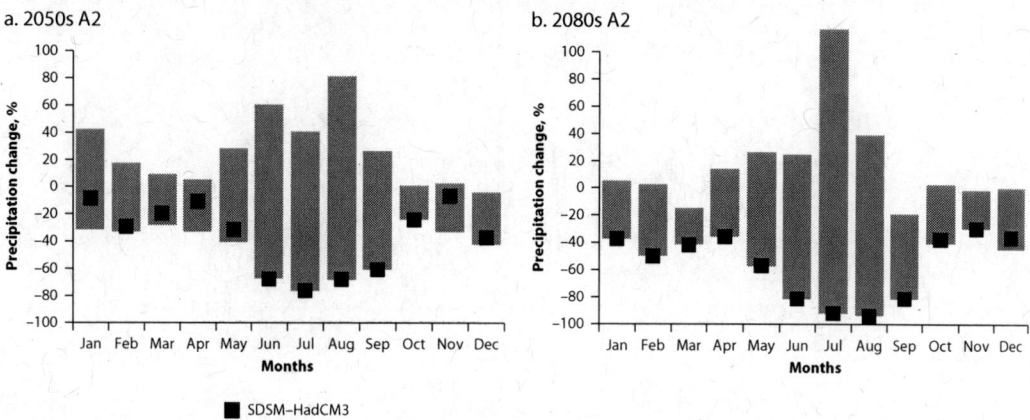

Source: Based on SDSM.

precipitation in some winter months up to the 2050s and in spring by the 2080s—highlighting the lack of consensus about future drying (figure 2.30). The SDSM and UCT scenarios are not directly comparable because of differences in the statistical techniques employed, choice of global climate model, and predictor variables used for the downscaling. These factors are all known to have a bearing on downscaled precipitation in the region, so it is good practice to employ a range of downscaled techniques (for example, Hertig and Jacobeit 2008; Wilby et al. 2009).

Olive Flowering

Although downscaling techniques can be used to simulate, infill (repair), and ensure the quality of observed meteorological data, downscaled scenarios are more often used for climate change impact assessments. However, given their inherent uncertainties, downscaled impacts should not be treated as predictions. They are best regarded as "narratives" of the future, and are most appropriately used in sensitivity testing and in appraising options for adaptation (Pielke and Wilby 2012).

A brief assessment of the potential impacts of climate change on olive flowering dates in northern Tunisia illustrates these points. The links between olive plant development (phenology) and local meteorological conditions in Tunisia are well documented (for example, Gargouri et al. 2008). It is generally accepted that growth does not occur below a 7°C temperature threshold in northern Tunisia and that 700–1,000 growing degree days (GDD) above this threshold are required for olives to flower fully. Depending on year-to-year weather conditions, full flowering typically occurs within 120–135 days from 1 January (Orlandi et al. 2010). Using these figures and the statistically downscaled

Figure 2.31 Annual Series of Simulated Full Flowering Date (Days from 1 January) Based on Downscaled TMAX under SRES A2 Emissions for a Site in North Tunisia

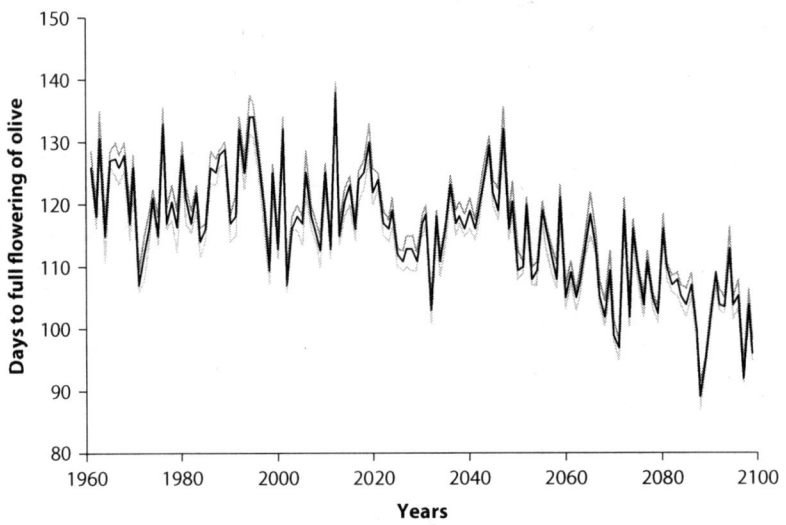

Source: Based on SDSM.

TMAX for Tunis, changes in the flowering date for olives were simulated for the period 1961–2100 (figure 2.31). An alternative modeling strategy is to take the present temperature series, adjust by the changes in the UCT ensemble, then compute the flowering dates for each run (figure 2.32).

Figure 2.32 Simulated Days from 1 January to Full Flowering of Olive under Present (1961–2000) and Future (2050s) TMAX Conditions at a Site in Northern Tunisia

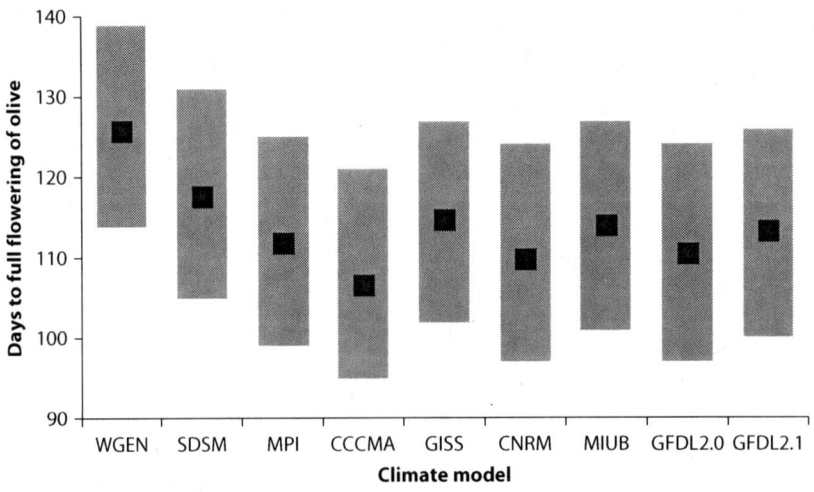

Source: World Bank data.
Note: WGEN is the simulated present; all other bars show the range and mean of the flowering dates for different climate models.

All of the climate scenarios show earlier flowering of olives due to expected increases in TMAX. SDSM flowering dates are on average 3 days earlier by the 2020s, 7 days earlier by the 2050s, and 16 days earlier by the 2080s (figure 2.31). However, there remains considerable variability in the number of days to flowering within these 30-year blocks. For example, by the 2050s, the simulated flowering dates could lie between April 8 and May 16.

The range of results from the ensemble of models also span about 25 days, but the mean and earliest flowering dates depend on the individual climate model (figure 2.32). The downscaled model with the greatest increase in January-May maximum temperatures (CCCMA) suggests flowering as early as the first week of April (compared with the last week of April at present). Nonetheless, a significant proportion of the flowering dates in the 2050s still fall within the present range. The ensemble results also do not include other factors that could affect future olive growth such as frost, drought, or disease, nor do they reflect possible changes in technology or cultivar.

Greater Scarcity of Water and Harm to Natural Systems

According to Gaaloul (2008, 240), Tunisia faces climatic challenges because of irregular and inadequate rainfall, a fragile ecosystem, limited natural resources, and the risk of over-exploitation of these resources. Not surprisingly, surface and groundwater resources, as well as agriculture and land, have been favored topics for climate impact research. Given Tunisia's nearly 2,000 kilometers of coastline and 83 square kilometers of Wetlands of International Importance (37 RAMSAR sites),[14] sea level rise also potentially has serious consequences for the country.

Surface and Groundwater

Even without climate change, Tunisia has scarce water resources. The national per capita sustainable water availability is roughly 400 cubic meters per year—well below both the average for the Middle East and North Africa (1,250 cubic meters per year) and the United Nations threshold for classifying regions as water scarce (1,000 cubic meters per year). When assessed by water availability-to-withdrawals, by consumption-to-dry flow volumes, or by per capita water availability, Tunisia emerges as a country that could suffer from severe water stress by the 2050s (Alcamo, Flörke, and Märker 2007; Arnell 2004). Climate model scenarios for North Africa consistently show reduced rainfall and runoff (for example Milly, Dunne, and Vecchia 2005; Vörösmarty, McIntryre, and Gessner 2010; World Bank 2011). However, semi-arid climate zones with annual rainfall of 400–600 millimeters are particularly vulnerable to reduced surface runoff, even from modest changes in rainfall. For example, an area with 500 millimeters per year rainfall, such as the vicinity of Kélibia, could lose 50 percent of perennial drainage density with just a 10 percent reduction in rainfall (de Wit and Stankiewicz 2006).

Several studies have investigated climate change impacts in Tunisia at the river basin scale. For example, Abouabdillah et al. (2010) statistically down-scaled precipitation and temperature from the Canadian Global Coupled Model (CGCM3.1) to the Merguellil basin (central Tunisia) and used the scenarios in the Soil and Water Assessment Tool (SWAT) to assess impacts on water balance components. Summer droughts were shown to intensify but variations in total water yield were also found to depend on altitude. Leduc et al. (2007) considered human impacts in the same basin (for example rainwater harvesting, groundwater pumping, and impoundment) on recharge and water levels in the Kairouan aquifer. Others have investigated the effects of traditional water harvesting system (such as *tabia*[15]) on hillside runoff (figure 2.33). Questions remain about the upper limit of rainfall that can be harvested using existing systems in central Tunisia (Nasri et al. 2004).

Groundwater has underpinned economic development, as well as improved quality and reliability of supplies—particularly during droughts—for rural populations across large parts of Tunisia (Besbes et al. 2010; Gaaloul 2008). Because coastal aquifers are highly vulnerable to overexploitation and possible saltwater intrusion, rising sea levels in the Mediterranean are a cause for concern in managing them. For example, following detailed modeling Zghibi, Zouhri,

Figure 2.33 Relationship Between Harvested Water and Rainfall over a *Tabia* System in the Bouhedma Catchment East of Gafsa

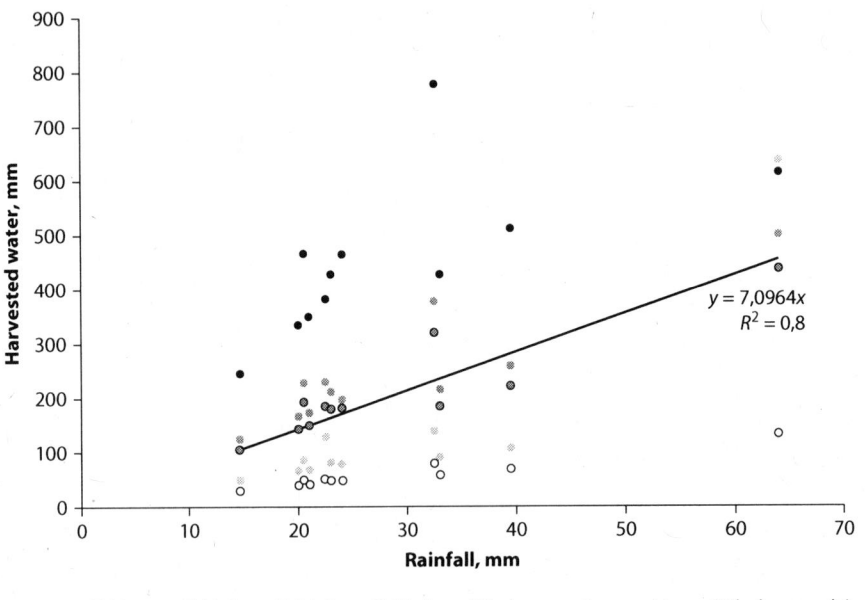

Source: Nasri et al. 2004, p. 270, Figure 6. Reprinted by permission of the publisher (Taylor & Francis Ltd., http://www.tandf.co.uk/journals).

and Tarhouni (2011) recommended a program of artificial recharge to retard the fall in the water table and hence control saltwater intrusion into the Cap-Bon peninsula's Korba aquifer.

Only 9 percent of Tunisia's renewable water resources originate outside the country, making it much less dependent on foreign sources than some other Arab countries (Medany 2008). Nonetheless, there is recognition of the need for greater cooperation and coordination in managing transboundary water resources as part of a broader portfolio of adaptation measures (Smith 1996; Scheumann and Alker 2009). This regimen includes establishing local water conservation, artificial recharge measures, and institutional frameworks to enable Algeria, Libya, and Tunisia to cooperatively manage the North West Sahara Aquifer System (NWSAS) in a changed climate context (Al-Gama 2011; Fezzani et al. 2005).

Agriculture and Land

With 93 percent of Tunisia's cropland dedicated to rainfed agriculture, the country's rural population is highly vulnerable to rainfall variability and the impact of long droughts (Mougou et al. 2008, 2011). Climate variability and change are also linked to evolving plant distributions that could have implications for pasture and soil conservation (for example, Belluci et al. 2011). Consequently, like water resources, agriculture and land management are the focus of a growing number of climate impact studies.

For example, Giannakopoulos et al. (2009) analyzed output from the HadCM3 model from the point of view of impact-relevant metrics for crop productivity and fire risk using grid cells across the Mediterranean, including one for Tunisia. (In fact, this study used the same climate model information as the above SDSM case studies.) Five cultivars were modeled (maize, sunflower, beans, potatoes, and wheat) for the period 2031–60, when a global warming of 2°C is assumed to occur. The change in crop productivity showed mixed results for Tunisia: maize and bean yields declined by 5–15 percent; sunflower and wheat increased by 5–15 percent; and potato yields changed by ±10 percent. Rainfed crops sown in the spring were adversely affected by the warmer and drier climate, but increases in carbon dioxide concentrations helped to offset some of the losses in yield. The authors advocate earlier sowing dates (to counteract a growing season shortened by higher temperatures and water stress) and use of cultivars that take longer to grow. However, both adaptation measures would be contingent on access to additional water for irrigation.

Because North African countries already import significant quantities of grain, there is considerable interest in the potential impacts of climate change on Tunisia's wheat crop. Lhomme, Mougou, and Mansour (2009) used scenarios from the ARPEGE climate model to simulate changes in durum wheat yields in Jendouba (northern Tunisia) and Kairouan (central Tunisia). The climate model shows reduced annual rainfall in Jendouba but an increase in Kairouan. As a consequence of these rainfall scenarios and higher temperatures, the soil moisture needed for sowing occurs earlier and growing cycles decrease in both locations. Gasmi, Belloumi, and Matoussi (2011) evaluated impacts on durum wheat in Béja and El Kef, two districts of northwest Tunisia. Rather than use climate model output, the authors used historic rainfall figures and increasing temperatures in the range 0.5–3.5°C, with increments of 0.5°C every 15 years in their econometric model. Inspection of the empirical model coefficients suggests that none of the meteorological drivers were statistically significant predictors of yield, so the results are difficult to interpret.

Potential increases in extreme rainfall could have consequences for soil erosion and siltation of reservoirs, particularly on slopes prone to high runoff (Ayadi et al. 2010). Depending on the level of soil degradation, average annual soil loss ranges between 14.5 and 36.4 tons per hectare in the Tunisian Dorsal, the eastern edge of the Atlas Mountains. The lifespan of reservoirs in the region is already less than 15 years (Jebari et al. 2010). Climate conditions also affect cropped soil salinity in irrigated lands through changes in infiltration, percolation to the shallow groundwater, soil evaporation, crop transpiration, groundwater flow, capillary rise, and drain outflow (Askri, Bouhlila, and Job 2010). Reduced irrigation (due to restrictions caused by increased water scarcity) and higher evaporation could increase soil salinity during the spring and summer seasons. Selective breeding of landraces (for instance, forage crops such as lucerne) could be one way of contending with salt-stressed environments or rainfed and irrigated areas receiving less than 350 millimeters of water in the spring and summer (Annicchiarico et al. 2011).

Coastal Zones

Dasgupta et al. (2007, 2009) evaluated how sensitive developing countries are to rises in sea level and tidal surges 1–5 meters higher than they are now. Their motivation was to develop an objective method to allocate aid according to the degree of threat (Dasgupta et al. 2007, 44). Based on six factors (land area, population, agriculture, urban extent, wetlands, and gross domestic product [GDP]) they identified Vietnam, the Arab Republic of Egypt, and the Bahamas as the World's most vulnerable countries. The cities found to be most at risk from more intense storm surges (by World Bank Group region) were Hai Phong (Vietnam), Barisal (Bangladesh), Bugama (Nigeria), Ciudad del Carmen (Mexico), and Port Said (Egypt).

Figure 2.34 shows that approximately 5 percent of the population and urban extent of Tunisia would be affected by a 1-metre sea level rise (Dasgupta et al. 2007). However, given the significance of coastal tourism such a rise could have a disproportionately large impact on the economy. A 1 in 100 year storm surge (that is, a particularly extreme event) superimposed on this amount of sea level rise would place 64 percent of Tunisia's coastal wetland at risk from inundation. This compares with a global average of about 30 percent for exposed wetlands in developing countries.

Figure 2.34 The Impact (Percent) of a 1-Meter Sea Level Rise on Country Area, GDP, Agricultural Extent, Urban Extent, and Wetland Extent in Arab Countries

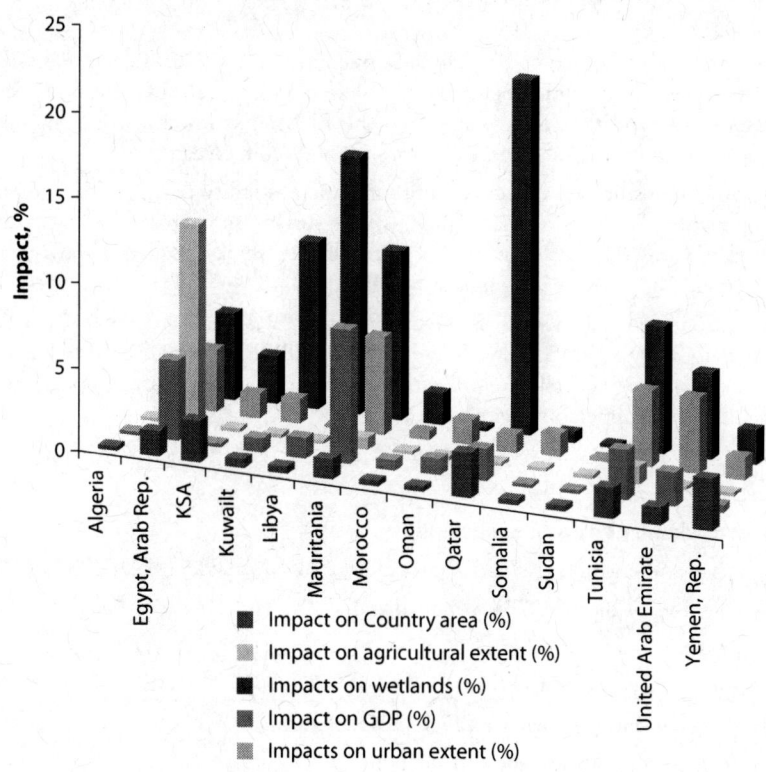

Source: Medany 2008, Figure 4 adapted from Dasgupta et al. 2007.

Given Tunisia's generally unfavorable climate change scenarios and the country's vulnerable rural populations, it makes sense to identify development strategies that work well in the wide range of conditions that the country faces now and potentially in the future. For instance, "no regret" strategies yield benefits regardless of climate change. In practice, there are always some opportunity costs, trade-offs, or externalities associated with such strategies, so it is perhaps better to view them as "low regret." These strategies should not only take into account present development priorities but also keep open options for further adaptation. For example, using water more efficiently and protecting it from salinization are rational measures to take no matter which climate change scenario pertains. Likewise, long-term monitoring and reporting on environmental quality are needed to track climatic and nonclimatic pressures, as well as to judge the success of any interventions. Cooperation and management of transboundary water resources (such as the NWSAS) are also priorities.

Notes

1. According to public meteorological records, derived climate indices, and satellite readings,

2. This variability is linked to two climatic phenomena: the North Atlantic oscillation (fluctuations in the difference of atmospheric pressure at sea level between the Icelandic low and the Azores high) and the El Niño southern oscillation (periodic surface warming/high air surface pressure in the Pacific Ocean).

3. These scenarios do not reflect the full range of possibility because other important components of regional climate (such as land surface and aerosol feedback) are not included. Consequently, even though most climate change scenarios show warming and drying, they cannot be considered predictive.

4. International efforts such as the World Climate Research Program's Coordinated Regional Climate Downscaling Experiment (CORDEX) aim to better characterize uncertainty in regional climate projections by prioritizing and comparing downscaling techniques for vulnerable regions including North Africa.

5. http://www.ncdc.noaa.gov/oa/gsod.html.

6. *Direction Générale des Resources en Eau.*

7. http://eca.knmi.nl/dailydata/index.php.

8. http://badc.nerc.ac.uk/view/badc.nerc.ac.uk__ATOM__dataent_1256223773328276.

9. http://cera-www.dkrz.de/CERA/index.html.

10. http://www.climatewizard.org/.

11. http://www.cccsn.ca/index-e.html.

12. http://wcrp.ipsl.jussieu.fr/SF_RCD_CORDEX.html.

13. http://cip.csag.uct.ac.za/webclient/introduction.

14. http://www.ramsar.org/cda/en/ramsar-documents-list-annotated-ramsar-15846/main/ramsar/1-31-218%5E15846_4000_0__.

15. A traditional water harvesting system consisting of two-thirds runoff collection area (used for grazing) and one-third cropped area arranged in a cascading series of U-shaped soil banks.

References

Abouabdillah, A., O. Oueslati, A. M. De Girolamo, and A. Lo Porto. 2010. "Modelling the Impact of Climate Change in a Mediterranean Catchment (Merguelli, Tunisia)." *Fresenius Environmental Bulletin* 19: 2334–47.

Alcamo, J., M. Flörke, and M. Märker. 2007. "Future Long-term Changes in Global Water Resources Driven by Socioeconomic and Climatic Changes." *Hydrological Sciences Journal* 52: 247–75.

Alexander, L. V., X. Zhang, T. C. Peterson, J. Caesar, B. Gleason, A. M. G. Klein Tank, M. Haylock, D. Collins, B. Trewin, F. Rahimzadeh, A. Tagipour, K. Rupa Kumar, J. Revadekar, G. Griffiths, L. Vincent, D. B. Stephenson, J. Burn, E. Aguilar, M. Brunet, M. Taylor, M. New, P. Zhai, M. Rusticucci, and J. L. Vazquez-Aguirre. 2006. "Global Observed Changes in Daily Climate Extremes of Temperature and Precipitation." *Journal of Geophysical Research* 111: D05109.

Al-Gama, S. A. 2011. "An Assessment of the Recharge Possibility to North-Wesyern Sahara Aquifer System (NWSAS) Using Environmental Isotopes." *Journal of Hydrology* 398: 184–90.

Annicchiarico, P., L. Pecetti, A. Abdelguerfi, A. Bouizgaren, A. M. Carroni, T. Hayek, M. M'Hammadi Bouzina, and M. Mezni. 2011. "Adaptation of Landrace and Variety Germplasm and Selection Strategies for Lucerne in the Mediterranean Basin." *Field Crops Research* 120: 283–91.

Arnell, N. W. 2004. "Climate Change and Global Water Resources: SRES Emissions and Socioeconomic Scenarios." *Global Environmental Change* 14: 31–52.

Askri, B., R. Bouhlila, and J. O. Job. 2010. "Development and Application of a Conceptual Hydrologic Model to Predict Soil Salinity within Modern Tunisian Oases." *Journal of Hydrology* 380: 45–61.

Ayache, F., J. R. Thompson, R. J. Flower, A. Boujarra, F. Rouatbi, and H. Makina. 2009. "Environmental Characteristics, Landscape History and Pressures on Three Coastal Lagoons in the Southern Mediterranean Region: Merja Zerga (Morocco), Ghar El Melh (Tunisia) and Lake Manzala (Egypt)." *Hydrobiologia* 622: 15–43.

Ayadi, I., H. Abida, Y. Djebbar, and M. Raouf Mahjoub. 2010. "Sediment Yield Variability in Central Tunisia: A Quantitative Analysis of Its Controlling Factors." *Hydrological Sciences Journal* 55: 446–58.

Bargaoui, Z., Y. Tramblay, E. Lawin, and E. Servat. Forthcoming. "Seasonal Precipitation Variability in Regional Climate Simulations Over Northern Basins of Tunisia." *International Journal of Climatology*.

Belluci, E., L. Nanni, E. Bitocchi, and M. Rossi. 2011. "Genetic Diversity and Geographic Differentiation in the Alternative Legume Tripodion Tetraphyllum (L.) Fourr. in North African Populations." *Plant Biology* 13: 381–90.

Bengtsson, L., K. I. Hodges, and E. Roeckner. 2006. "Storm Tracks and Climate Change." *Journal of Climate* 19: 3518–43.

Berndtsson, R. 1989. "Topographical and Coastal Influence on Spatial Precipitation Patterns in Tunisia." *International Journal of Climatology* 9: 357–69.

Berndtsson, R., and J. Niemczynowicz. 1986. "Spatial and Temporal Characteristics of High-intensive Rainfall in Northern Tunisia." *Journal of Hydrology* 87: 285–98.

Besbes, M., J. Chahed, A. Hamdane, and G. De Marsily. 2010. "Changing Water Resources and Food Supply in Arid Zones: Tunisia. In *Water and Sustainability in Arid Regions*, edited by G. Schneier Madanes and M. F. Courel. Proceedings of the First WATARD International Conference on Water, Ecosystems and Sustainable Development in Arid and Semi-Arid Areas, University Xinjiang, China.

Calafat, F. M., and D. Gomis. 2009. "Reconstruction of Mediterranean Sea Levels Fields for the Period 1945–2000." *Global and Planetary Change* 66: 225–34.

Calafat, F. M., M. Marcos, and D. Gomis. 2010. "Mass Contribution to Mediterranean Sea Level Variability for the Period 1948–2000." *Global and Planetary Change* 73: 193–201.

Christensen, J. H., B. Hewitson, A. Busuioc, A. Chen, X. Gao, I. Held, R. Jones, R. K. Kolli, W.-T. Kwon, R. Laprise, V. Magaña Rueda, L. Mearns, C. G. Menéndez, J. Räisänen, A. Rinke, A. Sarr, and P. Whetton. 2007. "Regional Climate Projections." In *Climate Change 2007: The Physical Science Basis*, edited by S. Solomon, D. Qin, M. Manning, Z. Chen, M. Marquis, K. B. Averyt, M. Tignor, and H. L. Miller. Contribution of

Working Group I to the Fourth Assessment Report of the Intergovernmental Panel on Climate Change. Cambridge: Cambridge University Press.

Corte-Real, J., X. Zhang, and X. Wang. 1995. "Large-scale Circulation Regimes and Surface Climatic Anomalies Over the Mediterranean." *International Journal of Climatology* 15: 1135–50.

Dai, A., and T. M. L. Wigley. 2000. "Global Patterns of ENSO-induced Precipitation." *Geophysical Research Letters* 27: 1283–86.

Dasgupta, S., B. Laplante, C. Meisner, D. Wheeler, and J. Yan. 2007. "The Impact of Sea Level Rise on Developing Countries: A Comparative Analysis." World Bank Policy Research Working Paper 4136, World Bank, Washington, DC, 51pp.

Dasgupta, S., B. Laplante, S. Murray, and D. Wheeler. 2009. "Sea Level Rise and Storm Surges: A Comparative Analysis of Impacts in Developing Countries." World Bank Policy Research Working Paper 4901, World Bank, Washington, DC, 43pp.

de Wit, M., and J. Stankiewicz. 2006. "Changes in Surface Water Supply Across Africa with Predicted Climate Change." *Science* 311 (5769): 1917–21.

Fezzani, C., D. Latrech, A. Mamou, and A. Trux. 2005. "The North-Western Sahara Aquifer System (NWSAS): Joint Management of a Transboundary Water Basin." *Agriculture and Rural Development* 1: 57–59. http://www.rural21.com/uploads/media/ELR_The_North-Western_Sahara_Aquifer_System_0105.pdf.

Gaaloul, N. 2008. "The Role of Groundwater During Drought in Tunisia." In *Climatic Changes and Water Resources in the Middle East and North Africa*, edited by F. Zereini and H. Hötzl, 552pp. Chapter 3.3. Berlin: Springer.

Gao, X., and F. Girogi. 2008. "Increased Aridity in the Mediterranean Region Under Greenhouse Gas Forcing Estimated from High Resolution Simulations with a Regional Climate Model." *Global Planetary Change* 62: 195–209.

Gargouri, K., A. Rhouma, A. Sahnoun, M. Ghribi, H. Bentaher, B. Ben Rouina, and M. Ghrab. 2008. "Assessment of the Impact of Climate Change on Olive Growing in Tunisia Using GIS Tools." *Options Méditerranéennes Series A* 80: 349–52.

Gasmi, F., M. Belloumi, and M. S. Matoussi. 2011. "Climate Change Impacts on Wheat Yields in the North-West of Tunisia." Economic Research Forum Working Paper 652, Giza, Egypt, 17pp.

Giannakopoulos, C., P. Le Sager, M. Bindi, M. Moriondo, E. Kostopoulou, and C. M. Goodess. 2009. "Climate Changes and Associated Impacts in the Mediterranean Resulting from a 2°C Global Warming." *Global and Planetary Change* 68: 209–24.

Giorgi, F., and X. Bi. 2005. "Updated Regional Precipitation and Temperature Changes for the 21st Century from Ensembles of Recent AOGCM Simulations." *Geophysical Research Letters* 32: L21715.

Giorgi, F., and P. Lionello. 2008. "Climate Change Projections for the Mediterranean Region." *Global and Planetary Change* 63: 90–104.

Haylock, M. R., N. Hofstra, A. M. G. Klein Tank, E. J. Klok, P. D. Jones, and M. New. 2008. "A European Daily High-Resolution Gridded Data Set of Surface Temperature and Precipitation for 1950–2006." *Journal of Geophysical Research* 113: D20119.

Hertig, E., and J. Jacobeit. 2008. "Assessments of Mediterranean Precipitation Changes for the 21st Century Using Statistical Downscaling Techniques." *International Journal of Climatology* 28: 1025–45.

Hewitson, B. C., and R. G. Crane. 2006. "Consensus between GCM Climate Change Projections with Empirical Downscaling: Precipitation Downscaling Over South Africa." *International Journal of Climatology* 26: 1315–37.

Hewitson, B. C., and R. L. Wilby. 2008. "A Climate Scenarios Portal for the MENA Region." Phase I Report on behalf of the World Bank, Washington, DC, 53pp.

Huffman, G. J., R. F. Adler, D. T. Bolvin, G. Gu, E. J. Nelkin, K. P. Bowman, Y. Hong, E. F. Stocker, and D. B. Wolff. 2007. "The TRMM Multi-Satellite Precipitation Analysis: Quasi-Global, Multi-Year, Combined-Sensor Precipitation Estimates at Fine Scale." *Journal of Hydrometeorology* 8: 38–55.

Jebari, S., R. Berndtsson, A. Bahri, and M. Boufaroua. 2010. "Spatial Soil Loss Risk and Reservoir Siltation in Semi-Arid Tunisia." *Hydrological Sciences Journal* 55: 121–37.

Kingumbi, A., Z. Bargaouti, and P. Hubert. 2005. "Investigation of the Rainfall Variability in Central Tunisia." *Hydrological Sciences Journal* 50: 493–508.

Klein, M., and M. Lichter. 2009. "Statistical Analysis of Recent Mediterranean Sea-Level Data." *Geomorphology* 107: 3–9.

Kouroutzoglou, J., H. A. Flocas, I. Simmonds, K. Keay, and M. Hatzaki. 2011. "Assessing Characteristics of Mediterranean Explosive Cyclones for Different Data Resolution." *Theoretical and Applied Climatology* 105: 263–75.

Leduc, C., S. Ben Ammar, G. Favreau, R. Beji, R. Virrion, G. Lacombe, J. Tarhouni, C. Aouadi, B. Zenati Chelli, N. Jebnoun, M. Oi, J. L. Michelot, and K. Zouari. 2007. "Impacts of Hydrological Changes in the Mediterranean Zone: Environmental Modifications and Rural Development in the Merguellil Catchment, Central Tunisia." *Hydrological Sciences Journal* 52: 1162–78.

Lhomme, J. P., R. Mougou, and M. Mansour. 2009. "Potential Impact of Climate Change on Durum Wheat Cropping in Tunisia." *Climatic Change* 96: 549–64.

Lionello, P., and F. Giorgi. 2007. "Winter Precipitation and Cyclones in the Mediterranean Region: Future Climate Scenarios in a Regional Simulation." *Advances in Geosciences* 12: 153–58.

Mangiarotti, S. 2007. "Coastal Sea Level Trends from TOPEX Poseidon Satellite Altimetry and Tide Gauge Data in the Mediterranean Sea during the 1990s." *Geophysical Journal International* 170: 132–44.

Marcos, M., and M. N. Tsimplis. 2007. "Forcing of Coastal Sea Level Rise Patterns in the North Atlantic and the Mediterranean Sea." *Geophysical Research Letters* 18: L18604.

———. 2008. "Comparison of Results of AOGCMs in the Mediterranean Sea During the 21st Century." *Journal of Geophysical Research-Oceans* 113: C12028.

Mariotti, A. 2010. "Recent Changes in the Mediterranean Water Cycle: A Pathway Toward Long-Term Regional Hydroclimatic Change?" *Journal of Climate* 23: 1513–25.

Medany, M. 2008. "Impact of Climate Change on Arab Countries." Arab Environment: Future Challenges, Beirut. http://www.afedonline.org/afedreport/english/book9.pdf.

MedCLIVAR. 2004. "White Paper on Mediterranean Climate Variability and Predictability." Technical report. http://web.lmd.jussieu.fr/~li/gicc_medwater/bibliog-raphie/MedCLIVAR_WP.pdf.

Milly, P. C. D., K. A. Dunne, and A. V. Vecchia. 2005. "Global Pattern of Trends in Streamflow and Water Availability in a Changing Climate." *Nature* 438: 347–50.

Mitchell, T. D., and P. D. Jones. 2005. "An Improved Method of Constructing a Database of Monthly Climate Observations and Associated High-Resolution Grids." *International Journal of Climatology* 25: 693–712.

Mitchell, T. D., M. Hulme, and M. New. 2002. "Climate Data for Political Areas." *Area* 34: 109–12.

Mougou, R., A. Abou-Hadid, A. Iglesias, M. Medany, A. Nafti, R. Chetali, M. Mansour, and H. Eid. 2008. "Adapting Dryland and Irrigated Cereal Farming to Climate Change in Tunisia and Egypt." In *Climate Change and Adaptation*, edited by N. Leary, J. Adejuwon, V. Barros, I. Burton, J. Kulkarni, and R. Lasco. Chapter 10. London: International Institute for Environment and Development.

Mougou, R., M. Mansour, A. Iglesias, R. Z. Chebbi, and A. Battaglini. 2011. "Climate Change and Agricultural Vulnerability: A Case Study of Rain-Fed Wheat in Kairouan, Central Tunisia." *Regional Environmental Change* 11: S137–42.

Nasr, Z., H. Amohammed, R. Gafrej Lahache, C. Maag, and L. King. 2008. "Drought Modelling Under Climate Change in Tunisia During the 2020 and 2050 Periods." *Options Méditerranéennes Series A* 80: 365–69.

Nasri, S., J. Albergel, C. Cudennec, and R. Berndtsson. 2004. "Hydrological Processes in Macrocatchment Water Harvesting the Arid Region of Tunisia: The Traditional System of Tabias." *Hydrological Sciences Journal* 49: 261–72.

Orlandi, F., M. Msallem, T. Bonofiglio, A. Ben Dhiab, C. Sgromo, B. Romano, and M. Fornaciari. 2010. "Relationship Between Olive Flowering and Latitude in Two Mediterranean Countries (Italy and Tunisia)." *Theoretical and Applied Climatology* 102: 265–73.

Ouachani, R., Z. Bargaoui, and T. Ouarda. 2011. "Power of Teleconnection Patterns on Precipitation and Streamflow Variability of Upper Medjerda Basin." *International Journal of Climatology* 33 (1): 58–76. doi:10.1002/joc.3407.

Oueslatia, A. 1992. "Salt Marshes in Gulf-of-Gabes (Southeastern Tunisia)—Their Morphology and Recent Dynamics." *Journal of Coastal Research* 8: 727–33.

———. 1995. "The Evolution of Low Tunisian Coasts in Historical Times—From Progradation to Erosion and Salinization." *Quaternary International* 30: 41–47.

Pal, J. S., F. Giorgi, X. Bi, N. Elguindi, F. Solmon, X. Gao, S. A. Rauscher, R. Francisco, A. Zakey, J. Winter, M. Ashfaq, F. S. Syed, J. L. Bell, N. S. Diffenbaugh, J. Karmacharya, A. Konare, D. Martinez, R. P. da Rocha, L. C. Sloan, and A. L. Steiner. 2007. "RegCM3 and RegCNET: Regional Climate Modeling for the Developing World." *Bulletin of the American Meteorological Society* 88: 1395–409.

Pielke, R. A. Sr., and R. L. Wilby. 2012. "Regional Climate Downscaling—What's the Point?" *Eos* 93: 52–53.

Raible, C. C., B. Ziv, H. Saaroni, and M. Wild. 2010. "Winter Synoptic-Scale Variability Over the Mediterranean Basin Under Future Climate Conditions as Simulated by the ECHAM5." *Climate Dynamics* 35: 473–88.

Sâadaoui, M., and M. Sakka. 2007. "Pluviométrie en Méditerranée occidentale et Oscillation Nord Atlantique." *Publication de l'Association Internationale de Climatologie* 20: 17–22.

Scheumann, W., and M. Alker. 2009. "Cooperation on Africa's Transboundary Aquifers— Conceptual Ideas." *Hydrological Sciences Journal* 54: 793–802.

Shahin, M. 2007. *Water Resources and Hydrometeorology of the Arab Region.* Vol. 59. Water Science and Technology Library. The Netherlands: Springer, 586pp.

Slimani, M., C. Cudennec, and H. Feki. 2007. "Structure du gradient pluviométrique de la transition Méditerranée-Sahara en Tunisie: Determinants géographiques et saisonnalité." *Hydrological Sciences Journal* 52: 1088–102.

Smith, J. B. 1996. "Development of Adaptation Measures for Water Resources." *International Journal of Water Resources Development* 12: 151–63.

Tebaldi, C., K. Hayhoe, J. M. Arblaster, and G. A. Meehl. 2006. "Going to Extremes: An Intercomparison of Model-Simulated Historical and Future Change in Extreme Events." *Climatic Change* 79: 185–211.

Tsimplis, M. N., M. Marcos, and S. Somot. 2008. "21st Century Mediterranean Sea Level Rise: Steric and Atmospheric Pressure Contributions from a Regional Model." *Global and Planetary Change* 63: 105–11.

van der Linden, P., and J. F. B. Mitchell, eds. 2009. "ENSEMBLES: Climate Change and Its Impacts: Summary of Research and Results from the ENSEMBLES Project." Met Office Hadley Centre, Exeter, UK, 160pp.

Vargas-Yanez, M., F. Moya, M. C. García-Martínez, E. Tel, P. Zunino, F. Plaza, J. Salat, J. Pascual, J. L. López-Jurado, and M. Serra. 2010a. "Climate Change in the Western Mediterranean Sea 1900–2008." *Journal of Marine Systems* 82: 171–76.

Vargas-Yanez, M., M. J. Garcia, J. Salat, M. C. Garcia-Martinez, J. Pascual, and F. Moya. 2008. "Warming Trends and Decadal Variability in the Western Mediterranean Shelf." *Global and Planetary Change* 63: 177–84.

Vargas-Yanez, M., P. Zunino, A. Benali, M. Delpy, F. Pastre, F. Moya, M. C. García-Martínez, and E. Tel. 2010b. "How Much is the Western Mediterranean Really Warming and Salting?" *Journal of Geophysical Research-Oceans* 115: C04001.

Vörösmarty, C. J., P. B. McIntryre, and M. O. Gessner. 2010. "Global Threats to Human Water Security and River Biodiversity." *Nature* 467: 555–61.

Wilby, R. L. 2008. "Dealing with Uncertainties of Future Climate: The Special Challenge of Semi-Arid Regions." Proceedings of the Water Tribune: Climate Change and Water Extremes, Expo Zaragoza, Spain.

Wilby, R. L., C. W. Dawson, and E. M. Barrow. 2002. "SDSM—A Decision Support Tool for the Assessment of Regional Climate Change Impacts." *Environmental and Modelling Software* 17: 145–57.

Wilby, R. L., J. Troni, Y. Biot, L. Tedd, B. C. Hewitson, D. M. Smith, and R. T. Sutton. 2009. "A Review of Climate Risk Information for Adaptation and Development Planning." *International Journal of Climatology* 29: 1193–215.

World Bank. 2011. "Climate Variability and Change: A Basin Scale Indicator Approach to Understanding the Risk to Water Resources Development and Management." Water Papers, World Bank, Washington, DC, 139pp.

Zereini, F., and H. Hötzl, eds. 2008. *Climatic Changes and Water Resources in the Middle East and North Africa.* Berlin: Springer, 552pp.

Zghibi, A., L. Zouhri, and J. Tarhouni. 2011. "Groundwater Modelling and Marine Intrusion in the Semi-Arid Systems (Cap-Bon, Tunisia)." *Hydrological Processes* 25: 1822–36.

CHAPTER 3

Economic Impacts of Climate Change in Tunisia:
A Global and Local Perspective

Photograph by Dorte Verner

Chapter 2 elaborated the first stage of the climate change adaptation pyramid by looking at how climate change science informs the assessment of climate change risk in Tunisia. This chapter builds on the assessment of risks and opportunities by applying economic analysis to the climate change risks highlighted by the science in chapter 2 and provides an economic analysis of both risks and opportunities to address them.

Climate change affects countries' economies through a variety of channels. Rising temperatures and changes in rainfall patterns affect agricultural yields of both rainfed and irrigated crops. Unchecked sea level rise leads to losses in land, landscape, and infrastructure. A higher frequency of droughts may impair hydropower production, and an increase in floods can significantly raise public investment requirements for physical infrastructure (Garnaut 2008; Stern 2007; World Bank 2007; Yu, Alam, et al. 2010; Yu, Zhu, et al. 2010). Such sector-level impacts will have knock-on effects on other sectors and thus influence economic growth, food security, and household incomes.

The global economic effects of climate change also affect individual countries through changes in food supply, trade flows, and commodity prices (Breisinger et al. 2011; Nelson et al. 2010). For example, Nelson et al. (2009, 2010) project that global food prices are bound to increase substantially as a consequence of continued high global population growth, changing food consumption patterns, and climate change. Taking higher food prices into consideration is therefore important for any climate change impact assessment at the country level. Depending on the net import or export position of countries and the net producing and consuming status of households of specific affected commodities, the agricultural sector, household incomes, and food security are likely to be affected differently.

For Tunisia, both global and local climate change impacts are likely to matter for future development, given the country's relatively high levels of food import dependency. Tunisia imports between 50 and 88 percent of cereals and is a net importer of many other food items resulting in a moderate risk of food insecurity (Breisinger et al. 2012). Thus, climate change impacts add to the already significant development challenges that Tunisia is facing in light of the Arab awakening. Against this background, this chapter assesses how far climate change is likely to affect Tunisia and thus needs to be considered in future development strategies. It focuses on climate change impacts on agriculture and household welfare (taking economywide effects into consideration) and the effects of rising global food prices over 30 years as projected by IFPRI's International Model for Policy Analysis of Agricultural Commodities and Trade (IMPACT) model. The remainder of the chapter is structured as follows. Section 2 presents the analytical and empirical framework of the study and describes each of its components. Section 3 presents the results of the local, global, and combined climate change impact assessment, and Section 4 concludes.

Analytical and Empirical Framework

Global Impacts: IMPACT Model

The challenge of modeling climate change impacts arises in the wide-ranging nature of processes that underlie the working of markets, ecosystems, and human behavior.[1] The analytical framework used in this paper integrates various modeling components that range from the macro to the micro and from processes driven by economics to those that are essentially biophysical in nature. This section gives an overview of the model, data, and assumptions and box 3.1 summarizes some of the limitations of the modeling suite. More technical details can be found in Rosegrant et al. (2008) and Nelson et al. (2009, 2010).

The IMPACT model is a partial equilibrium agricultural model incorporating 32 crop and livestock commodities, including cereals, soybeans, roots and tubers, meats, milk, eggs, oilseeds, oilcakes and meals, sugar, and fruits and vegetables. IMPACT distinguishes 115 countries (or in a few cases country aggregates), within each of which supply, demand, and prices for agricultural commodities are determined. Large countries are further divided into major river basins. The results are called food production units. The model links the various countries and regions through international trade using a series of linear and nonlinear equations to approximate the underlying production and demand relationships. World agricultural commodity prices are determined annually at levels that clear international markets. Growth in crop production in each country is determined by crop and input prices, exogenous rates of productivity growth and area expansion, investment in irrigation, and water availability. Demand is a function of prices, income, and population growth and contains four categories of commodity demand: food, feed, biofuels feedstock, and other uses.

The IMPACT climate change modeling system combines a biophysical model (the Decision Support System for Agrotechnology Transfer [DSSAT] crop modeling suite, Jones et al. 2003) of responses of selected crops to climate, soil, and nutrients with the IFPRI Spatial Production Allocation Model dataset of crop location and management techniques (You and Wood 2006). These results are then aggregated and fed into IMPACT. For future climate, we use the Fourth Assessment Report of the United Nations Intergovernmental Panel on Climate Change (IPCC 2007) that runs using the Commonwealth Scientific and Industrial Research Organization (CSIRO) A1B and the Model for Interdisciplinary Research on Climate (MIROC) A1B models. For more information on the downscaling methodology, please refer to Breisinger et al. (2011). We assume that the resulting modeled crop yields change smoothly between their values in 2000 and 2050. This assumption eliminates any random extreme events such as droughts or high-rainfall periods. More precisely, such events are only part of the population of weather realizations leading to the modeled yields over which the average yield is computed. It also assumes that the forcing effects of greenhouse gas emissions proceed smoothly, that is, we do not see a gradual speedup in climate change. The effect of this assumption is to underestimate negative effects from climate variability.

Local Impacts: Impacts on Yields

Yield changes are determined for the seven major agroecological zones (AEZs) making up Tunisia. The projected yields come from simulations using crop models in the DSSAT crop modeling framework. The DSSAT crop simulation model is an extremely detailed, process-oriented model of the daily development of a crop, from planting to harvest ready (Jones et al. 2003). We considered three crops important to Tunisia: wheat, barley, and potatoes.[2] The DSSAT crop models are process-based crop simulation models. They require a large amount of input data but then can step through the prospective growing season on a daily basis and model how the plant grows, uses water and nutrients, responds to the weather, and ultimately accumulates mass in the harvested portion of the plant. This specificity makes the crop models a powerful tool for assessing the potential effects of climate change on crop yields at a very local geographic level, which can then be aggregated for use in the economic models.

The most important inputs for this application were the choice of planting dates and the climatic conditions. The planting dates were chosen as follows. Wheat was planted in the generally prevalent planting season of November. Determining the appropriate planting month for barley and potatoes was more difficult. These crops were planted in all months and the highest yielding month was chosen as the most likely. The initial soil conditions for all crops were set to contain a small amount of moisture and nitrogen as well as residues from previous production activities.

The climatic conditions were chosen to be consistent with those in the IMPACT world market price projections: baseline 2000 and 2050 climate as projected by CSIRO A1B and MIROC A1B downscalings from the Future Clim product (Jones, Thornton, and Heinke 2010). In general, the seasonal patterns of temperature and precipitation do not change much between the baseline and 2050 projections, so the same planting strategy was used for both cases. Of course, temperatures and rainfall amounts do change, resulting in sometimes dramatically different yields. Since the crop simulation models require daily weather data and the climate data are available as monthly averages, a random weather generator within the DSSAT framework (SIMMETEO) was used to create daily realizations consistent with monthly averages. For each location, 80 years of simulations were run using different weather for each one. The final yield for each location is the average across these 80 repetitions.

Once the yields were determined for each five-arcminute pixel in Tunisia, they were aggregated up to the AEZ level. The AEZ yields were computed as the area-weighted average yield. The projected yields for each pixel were multiplied by the production area thought to be present within that pixel. Aggregating across these provides the total production. Aggregating only the production areas provides the total area. Then the average yield is simply the total production divided by the total area. The production areas by crop within each pixel were assigned by using the maps from the Spatial Production Allocation Model (SPAM) (You et al. 2000, 2006).

Tunisia Dynamic Computable General Equilibrium Model

Climate change affects world prices and local agricultural productivities with direct implications for agriculture and indirect implications for processed food and the whole economy. We therefore develop a recursive dynamic computable general equilibrium (DCGE) model for Tunisia, which distinguishes several agricultural and processing sectors, while the rest of the industry sector and services is highly aggregated. A detailed description of the model structure and equations can be found in Thurlow (2004).

Producers in the model are price takers in output and input markets and maximize profits using constant returns to scale technologies. Primary factor demands are derived from constant elasticity of substitution (CES) value added functions, while intermediate input demand by commodity group is determined by a Leontief fixed-coefficient technology. The decision of producers between production for domestic and foreign markets is governed by constant elasticity of transformation (CET) functions that distinguish between exported and domestic goods in each traded commodity group in order to capture any quality-related differences between the two products. Under the small-country assumption, Tunisia faces perfectly elastic world demand curves for its exports at fixed world prices. On the demand side, imported and domestic goods are treated as imperfect substitutes in both final and intermediate demand under a CES Armington specification. Households use their incomes to consume commodities according to a linear expenditure system (LES) specification.

There are three labor categories in the model, with all types assumed to be fully employed and mobile across sectors. The assumption of full employment is consistent with widespread evidence that, while relatively few people have formal sector jobs, the large majority of working-age people engage in activities that contribute to gross domestic product (GDP). Capital is also assumed to be fully employed and mobile across sectors, reflecting the long-term perspective of this study. In agriculture, cultivated land, which is differentiated into rainfed, irrigated, and perennial land is sector specific, that is it cannot be reallocated across crops in response to shocks. Moreover, cropping decisions are made at the beginning of the period before the realization of climate shocks is imposed.

The DCGE model is based on a 2001 social accounting matrix built by Chemingui. It is specifically built to capture the economic and distributional effects of climate change in the Republic of Yemen. Given the importance of agriculture for income generation and the satisfaction of consumption needs, the model captures the sector of crop production and its linkages to other sectors such as food processing, manufacturing and services. The model includes 21 production activities and commodities, 7 factors of production, and 10 household types, distinguished by their income level. The 11 agricultural production activities are split into livestock (1), fishing (1), forestry (1), and crop production activities (8). Other production sectors and commodities included in the model are food processing (8), (other) industry (1), and services (1). The household groups are separated into income deciles but can be grouped into three household categories, according to their dominant source of functional income: (rural) farm households (deciles 1–3) receive more than 50 percent of their labor income from family labor and agricultural labor; in rural nonfarm households (deciles 4–6) nonagricultural labor income dominates; the upper 4 income deciles (deciles 7–10) receive no labor income (but capital and land income) from agriculture and are classified as urban households. This differentiation of household groups allows us to capture the distinctive patterns of income generation, and consumption as well as the distributional impacts of climate change.

The model runs from 2001 to 2030 and is recursive dynamic. Investments are savings driven, and savings grow proportionally to household income. Capital is fully employed and mobile to reflect the long-term perspective of the analysis. The Tunisian workforce is assumed to grow at an average long-term trend of 2 percent. Land is fixed, which means that current cultivated land cannot be expanded in the future. This assumption seems reasonable given the limited growth potential of the agricultural sector due to severe water constraints. Finally, total factor productivity (TFP) is assumed to grow at 1.7 percent annually for non agricultural sectors and at 0.75 percent annually for agricultural sectors over the period 2001–30. This two-speed TFP growth depicts the expected structural change under a business as usual scenario that is observed in many transforming countries (Breisinger and Diao 2008).

The model allows for some autonomous adaptation to climate change. Yield changes from the DSSAT model enter the production function of the DCGE model. These crop-specific and agro-ecological zone-specific changes in productivity change the returns to factors and alter output prices. For example, farm households can decide to employ their factors of production, such as labor, for nonfarm activities instead of growing crops and raising livestock. Or imported food can replace locally grown food when relative prices of locally grown food increase (and vice versa). A set of important elasticities guide these adjustments, including the substitution elasticity between primary inputs in the value-added production function, the elasticity between domestically produced and consumed goods and exported or imported goods, and the income elasticity in the demand functions. Income elasticities for Tunisia range from 0.3 for grains to 2.2 for services. For the factor substitution elasticity we choose 3.0, the elasticity of transformation is 4.0 and the Armington elasticity is 6.0 for all goods and services.

Based on the DCGE model we design three sets of scenarios. The first set of scenarios captures the global impacts of climate change, while the second set of scenarios assesses the local impacts of climate change. The third set combines the two to assess the joint effects (table 3.1). Within the first set of scenarios, we design three variants: scenario 1 changes the world food prices consistent with

Table 3.1 Climate Change Scenarios

Scenario	Change in model	Input
Baseline	See text	See text
Global impacts of climate change		
Scenario 1	Perfect mitigation, compared to base	IMPACT, Perfect mitigation
Scenario 1A	Climate change	IMPACT, MIROC A1B
Scenario 1B	Climate change	IMPACT, CSIRO A1B
Local impacts of climate change		
Scenario 2A	Crop yield changes	DSSAT MIROC A1B
Scenario 2B	Crop yield changes	DSSAT CSIRO A1B
Joint impacts of climate change		
Scenario 3A	1A and 2A	IMPACT and DSSAT, MIROC A1B
Scenario 3B	1B and 2B	IMPACT and DSSAT, CSIRO A1B

Source: World Bank data.

IMPACT model results under perfect mitigation of climate change. Scenario 1A explores climate change-related price effects under MIROC A1B, with the assumption that no climate change impacts are felt locally in Tunisia. Scenario 1B is a scenario to test the sensitivity of results to alternative price projections under CSIRO A1B. Scenario 2 imposes the yield changes from the DSSAT model on a crop by crop level. These include changes in yields for wheat, other grains (barley) and vegetables (potato) directly taken from the DSSAT model. In the absence of more specific information and consistent with the literature, we assume that yields of other agricultural crops such as fruits and olives are also negatively affected. Results for scenarios 1A–3B are reported as a change from the perfect global mitigation scenario to isolate the climate change effects.

Box 3.1

Limitations of the Modeling Suite

When interpreting the results, it is important to bear in mind some of the limitations of the modeling suite presented in this report. In total there are 22 Global Climate Models (GCMs); however, model output data are not available for all combinations of GCM and emissions scenarios, at least not the basic core variables for many crop and pasture models such as precipitation, maximum daily temperature, and minimum air temperature. This, in turn, severely restricted the choice of GCMs. From the CMIP3 dataset, Jones, Thornton, and Heinke (2009) used three GCMs—the National Center for Meteorological Research[3] Coupled Model Version 3 (CNRMCM3), The CSIRO Mark 3.0 (CSIRO-Mk3.0), and MIROC 3.2[4] (medium resolution)—and obtained maximum and minimum temperature data for the ECHam5[5]

(box continues on next page)

Box 3.1 Limitations of the Modeling Suite *(continued)*

model from another source (the Climate and Environmental Retrieval and Archive [CERA] database at DKRZ) for the three SRES scenarios. This restricted our choice of GCMs with sufficient data for crop modeling and we chose to focus on the results from MIROC and CSIRO projections. MIROC projects lower rainfall and higher temperatures than CSIRO, therefore, results from MIROC can be interpreted as a more pessimistic scenario and CSIRO as a more optimistic scenario. It is also important to note that the IMPACT model used for the global price analysis is a longer-term model whose results focus on climate change and not the shorter-run climate variability that may impact agricultural goods, and thus, short-term volatilities are not taken into consideration. For the crop modeling, we mainly rely on data from IFPRI's Spatial Allocation Model (SPAM) for Tunisia. Ideally crop yield projections and scenarios would exist for all major crops in Tunisia; however, due to the unavailability of detailed and consistent biophysical data and respective DSSAT crop models, we were restricted to several major crops namely wheat, barley and potatoes (both irrigated and rainfed). Livestock could not be considered, indicating that the impacts of climate change on agriculture may be underestimated. One major limitation of the DCGE model used is that while technological progress will likely be the key for successful adaptation, the model can only capture part of the adaptation process, such as shifts from one *existing* technology to another. However, it cannot capture potential future "breakthrough" technological advances that are not yet known. Such breakthroughs in technological advancements in any of the sectors are likely to have positive spillover effects into the rest of the sectors and household incomes.

Impacts of Climate Change on Global Food Prices and Local Yields

Before turning to the simulations, we briefly describe some structural characteristics of Tunisia's economy in order to set the stage for the evaluation of climate change impacts.

Structure of the Tunisian Economy

Agriculture is an important part of the Tunisian economy, accounting for 12–16 percent of GDP, depending on the size of the harvest. The sector provides jobs for about 20 percent of the country's labor force. The two most important export crops are cereals and olive oil, with almost half of all the cultivated land used for cereal production and another third planted with olive trees. Tunisia is one of the world's biggest producers and exporters of olive oil, and it also exports dates and citrus fruits. Tunisia remains one of the few Arab countries that produce most of the dairy products, vegetables, fruits, and red meat consumed domestically. Since the 1980s, agricultural output has increased by about 40 percent, and food exports have risen considerably.

Tunisia's labor-intensive agricultural sector uses very low levels of fertilizers and pesticides. Most of the land is split into small farms making production very inefficient. Some 80 percent of farms are smaller than 20 hectares, and only 3 percent are larger than 50 hectares. Severe droughts, such as the one experienced in 2000, have proven to be enormously costly. Tunisia is water scarce. Rainfall is characterized by its scarcity and spatial and temporal variability (Mougou, Rejeb, and Lebdi 2002). Variability and scarcity of water resources and high temperatures negatively affect the production in rainfed agriculture, especially cereals that are mainly produced under rainfed conditions (the large majority of total cereal area is rainfed).

Tunisia is also a net importer of major food items, including cereals, forage crops, and processed food. Overall agriculture's trade orientation is very low and uneven, with imports accounting for more than 15 percent of total domestic consumption and exports accounting for less than 5 percent of domestic production. Agriculture and related processing contribute about 17 percent to GDP. Food and agriculture-related processing make up about 30 percent of household consumption expenditures. Within this category, food processing constitutes the largest share of consumption, followed by fruits and vegetables (table 3.2).

Dividing up households according to socioeconomic characteristics, such as their location and occupation, allows for the analysis of income and

Table 3.2 Structure of the Tunisian Economy by Sector

	GDP share	Private consumption share	Export share	Export intensity	Import share	Import intensity
Wheat	1.2	0.1	0.1	3.7	1.9	47.5
Other cereals	0.2	0.1	—	—	2.1	87.8
Legumes	0.4	0.3	—	—	0.1	8.3
Forage crops	0.2	0.0	0.0	3.6	0.6	67.7
Olives	0.8	0.0	—	—	—	—
Other fruits	2.7	3.8	1.1	13.3	0.1	1.5
Vegetables	2.4	3.5	—	—	0.2	2.5
Other agriculture	0.1	0.1	0.1	16.9	0.1	32.8
Livestock	4.2	1.6	—	—	0.1	1.5
Forestry	0.3	0.3	—	—	0.0	7.4
Fishing	1.1	1.6	0.2	4.9	0.2	5.6
Meat	0.3	4.9	—	—	—	—
Milk and its products	0.6	2.1	—	—	0.2	9.5
Flour milling and its products	0.8	4.1	0.4	3.9	0.2	2.7
Oils	0.4	1.1	1.7	36.2	1.1	35.7
Canned food products	0.2	0.8	1.1	33.6	0.1	10.7
Sugar and its products	0.3	1.4	0.1	2.8	0.7	32.1
Other food products	0.5	4.9	0.7	5.8	0.8	12.1
Beverages	0.6	1.2	0.3	4.8	0.3	9.2
Other industries	29.7	33.2	79.8	35.3	85.6	46.2
Services	53.3	35.0	14.5	9.9	5.8	5.5
Total, of which:	—	—	100.0	21.2	100.0	28.4
Agriculture	13.4	11.3	1.5	3.5	5.4	16.4
Nonagriculture	86.6	88.7	98.5	23.0	94.6	29.7

Source: Tunisia DCGE model.

Note: — = not available.

Table 3.3 Household Income Sources by Income Type and Household

Households	Agricultural labor	Family labor	Nonagricultural labor	Capital	Rainfed land	Irrigated land	Perrenial land	Enterprises	ROW	Total	Population	Per-capita income
Decile1	97.0	250.0	111.7	8.1	0.1	0.2	0.2	0.1	7.1	474.5	945,600	501.8
Decile2	121.2	194.9	220.9	91.7	1.3	1.9	2.7	1.1	95.6	731.3	945,600	773.4
Decile3	143.4	151.4	321.5	168.6	2.4	3.5	5.0	2.1	165.5	963.4	945,600	1,018.8
Decile4	119.1	157.8	438.0	257.8	3.6	5.4	7.6	3.2	238.2	1,230.7	945,600	1,301.5
Decile5	—	120.3	554.6	347.0	4.9	7.3	10.2	4.3	450.8	1,499.4	945,600	1,585.7
Decile6	—	79.3	904.3	769.2	10.8	16.1	22.6	9.6	26.0	1,837.9	945,600	1,943.6
Decile7	—	—	1,083.3	906.2	12.8	19.0	26.6	11.3	226.3	2,285.5	945,600	2,417.0
Decile8	—	—	1,315.2	1,083.6	15.3	22.7	31.8	13.5	335.7	2,817.8	945,600	2,979.9
Decile9	—	—	1,789.4	1,446.4	20.4	30.3	42.5	18.0	579.7	3,926.7	945,600	4,152.6
Decile10	—	—	3,369.8	2,655.8	27.4	65.4	77.9	33.1	1,388.2	7,617.6	945,600	8,055.8
Farm	361.6	596.3	654.1	268.4	3.8	5.6	7.9	3.3	268.2	2,169.2	2,836,800	764.7
Nonfarm	119.1	357.4	1,896.9	1,374.0	19.3	28.8	40.4	17.1	715.0	4,568.0	2,836,800	1,610.3
Urban	—	—	7,557.7	6,092.0	75.9	137.4	178.8	75.9	2,529.9	16,647.6	3,782,400	4,401.3
Total rural	480.7	953.7	2,551.0	1,642.4	23.1	34.4	48.3	20.4	983.2	6,737.2	5,673,600	1,187.5
Total	480.7	953.7	10,108.7	7,734.4	99.0	171.8	227.1	96.3	3,513.1	23,384.8	9,456,000	2,473.0

Source: Tunisia DCGE model.
Note: — = not available.

Table 3.4 Household Income Sources by Income Type and Household Category in Tunisia, 2001 (Percent)

Households	Agricultural labor	Family labor	Nonagri-cultural labor	Capital	Rainfed land	Irrigated land	Perrenial land	Enterprises	ROW	Total
Decile1	20.4	52.7	23.5	1.7	0.02	0.04	0.04	0.02	1.5	100.0
Decile2	16.6	26.7	30.2	12.5	0.18	0.26	0.37	0.15	13.1	100.0
Decile3	14.9	15.7	33.4	17.5	0.25	0.36	0.52	0.22	17.2	100.0
Decile4	9.7	12.8	35.6	20.9	0.29	0.44	0.62	0.26	19.4	100.0
Decile5	—	8.0	37.0	23.1	0.33	0.49	0.68	0.29	30.1	100.0
Decile6	—	4.3	49.2	41.9	0.59	0.88	1.23	0.52	1.4	100.0
Decile7	—	—	47.4	39.6	0.56	0.83	1.16	0.49	9.9	100.0
Decile8	—	—	46.7	38.5	0.54	0.81	1.13	0.48	11.9	100.0
Decile9	—	—	45.6	36.8	0.52	0.77	1.08	0.46	14.8	100.0
Decile10	—	—	44.2	34.9	0.36	0.86	1.02	0.43	18.2	100.0
Farm	16.7	27.5	30.2	12.4	0.18	0.26	0.36	0.15	12.4	100.0
Nonfarm	2.6	7.8	41.5	30.1	0.42	0.63	0.88	0.37	15.7	100.0
Urban	—	—	45.4	36.6	0.46	0.83	1.07	0.46	15.2	100.0
Total rural	7.1	14.2	37.9	24.4	0.34	0.51	0.72	0.30	14.6	100.0
Total	2.1	4.1	43.2	33.1	0.42	0.73	0.97	0.41	15.0	100.0

Source: Tunisia DCGE model.

Note: — = not available.

distributional effects of climate change. Farm households, which make up 30 percent of the total population (according to our classification), earn about 9 percent of all household incomes, while the population and income shares are 30 and 19 percent for rural nonfarm households and 40 and 70 percent for urban households (see table 3.3).

By definition, household income levels are strongly related to factor and human capital endowments. Farm households receive most of their income from family and agricultural labor (each about 30 percent), while urban and rural nonfarm households rely on nonagricultural labor and capital (see table 3.4).

It is against these structural characteristics of the Tunisian economy and its households that the next sections analyze potential climate change impacts.

Global Climate Change Impacts

World food prices are projected to increase through demographic and income effects, which are augmented by climate change. Figure 3.1 reports the effects of the climate change scenarios of two global climate models on world food prices (CSIRO A1B and MIROC A1B). It also reports the price effects under perfect mitigation. With perfect mitigation, world prices for important agricultural crops such as wheat and maize will increase between 2000 and 2030 under both scenarios, driven by population and income growth and biofuels demand. The price of maize and wheat is projected to rise by 63 percent and 39 percent, respectively. Climate change results in additional price increases: a total of 52–55 percent for maize and 94–111 percent for wheat (Nelson et al. 2009).[6] Prices of vegetables and fruits as well as cotton hardly change over time in the

Figure 3.1 Global Food Price Scenarios

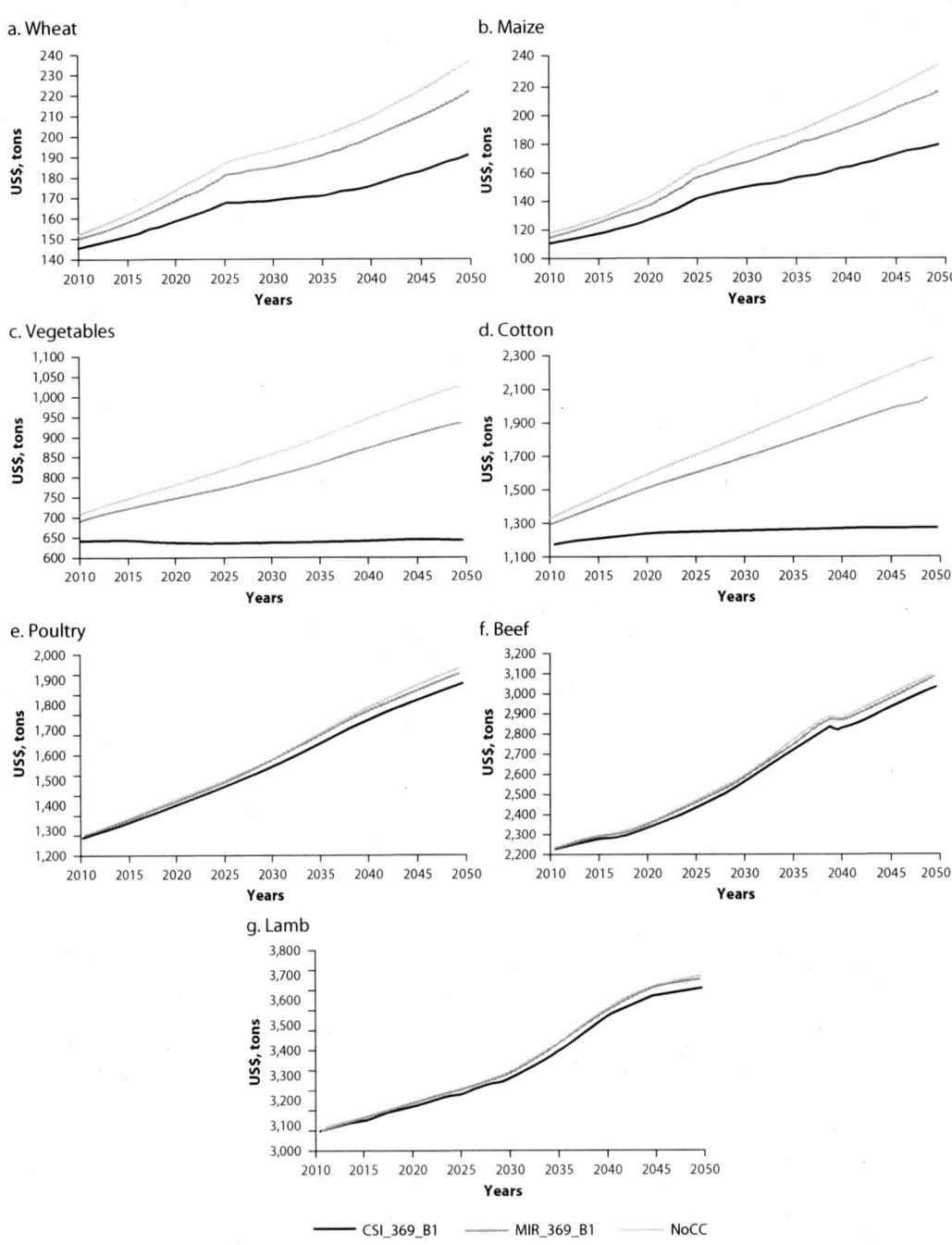

a. Wheat

b. Maize

c. Vegetables

d. Cotton

e. Poultry

f. Beef

g. Lamb

CSI_369_B1   MIR_369_B1   NoCC

Source: IFPRI's IMPACT model.
Note: NoCC stands for no climate change or perfect mitigation. Tons are in metric tons. The sudden changes for wheat and maize in 2024 reflect assumptions made on the phase out of biofuel policies. We eliminate this policy driven change when implementing the price changes in the computable general equilibrium (CGE) model by assuming that the price trend of 2025–30 follows the same trend as between 2025 and 2020.

perfect mitigation scenario but are expected to rise considerably as a consequence of climate change. Livestock are not directly affected by climate change in IMPACT. However, the effects of higher feed prices caused by climate change pass through to livestock, resulting in somewhat higher meat prices.

Local Climate Change Impacts

Temperature and Rainfall Variations in Tunisia

Results from the spatially downscaled climate projections show that temperatures are expected to rise over their baseline counterpart under both the CSIRO and the MIROC Global Climate Model (GCM) scenarios. However, the variation in temperatures over their baseline equivalents—both minimum and maximum—differs under the CSIRO and the MIROC scenarios (figure 3.2). The temperature variations under the MIROC scenario are the highest, followed by the temperature variations projected under the CSIRO scenario. In August, the MIROC monthly maximum temperatures rise 3.4°C above the baseline maximum temperatures for that month and rise 3.1°C above the baseline for the average monthly temperatures for July. Under the MIROC scenario, the variations are far greater for both the minimum and maximum temperatures. Over the course of the year, the MIROC scenario projects a more than 2.2°C rise in temperatures by 2050 in minimum temperatures over the baseline, and from June to September projects that minimum temperatures will rise over their baseline values by more than 3.2°C. Maximum temperatures are also expected to increase over their baseline values under the MIROC scenario. For

Figure 3.2 Average Monthly Temperature in Tunisia (°C)

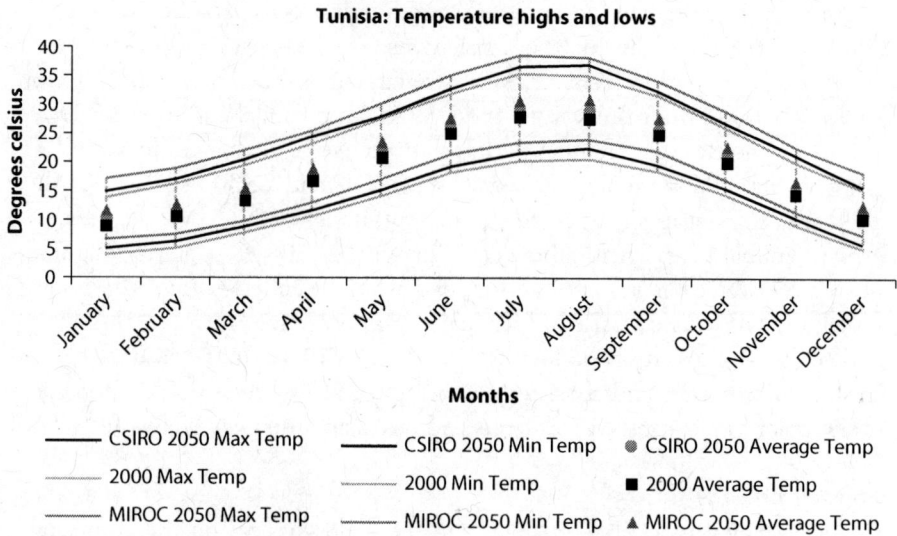

Source: Based on Jones, Thornton, and Heinke 2010.

Figure 3.3 Average Monthly Rainfall (Millimeters)

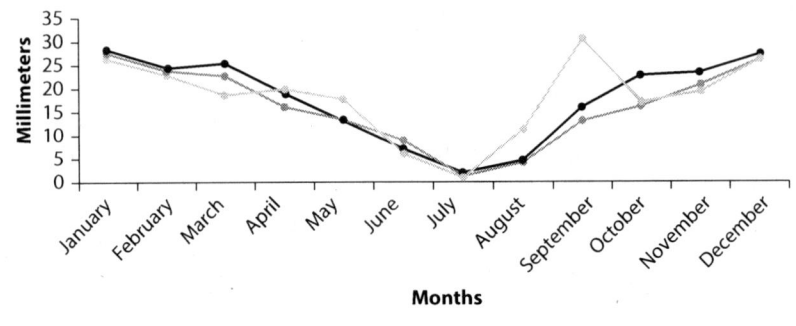

Tunisia: Average monthly rain

Source: Based on Jones, Thornton, and Heinke 2010.

the entire year, the MIROC maximum temperatures rise by more than 2°C and for four months out of the year, MIROC temperature highs are expected to rise more than 3°C over their baseline equivalents.

Variation in average monthly rainfall across Tunisia, as predicted by the CSIRO and MIROC GCM scenarios, is relatively higher only for the latter scenario. As figure 3.3 shows, average monthly rainfall (in millimeters) of the CSIRO scenario roughly follows the baseline except from September to November where it falls slightly below.

However, the MIROC scenario projects an increase in rainfall[7] from June to October across Tunisia. From October to December, rainfall under the MIROC scenario is slightly below that projected under the baseline.

Changes in rainfall and temperature are the main drivers of yield changes: all else was kept the same for the simulations. Yield changes over time, due to climate change, are projected to vary strongly across the three selected crops. Table 3.5 shows the results from the DSSAT crop model for Tunisia.[8] Driven mainly by the diverging rainfall patterns, projected yield changes for wheat and potatoes differ substantially between the MIROC and CSIRO scenarios. Wheat and barley are negatively affected throughout, under both GCM scenarios, with average annual wheat yield falling more under the MIROC scenario than under the CSIRO for both irrigated and rainfed wheat. Rainfed potatoes fare better under the MIROC scenario, which projects more rain. However, the average annual yield for irrigated potatoes is lower under MIROC. In the absence of more specific information, and consistent with the literature, we assume that yields of other agricultural crops such as fruits and olives are also negatively affected.

Impacts on GDP

Imposing global food price changes and the yield changes on the computable general equilibrium (CGE) model allows for assessing the economywide effects of climate change on the Tunisian economy and households. Results show that

Table 3.5 Average Annual Yield Changes for Selected Crops, 2000–50

Crop	MIROC (% Yield changes)		CSIRO (% Yield changes)	
	Irrigated	Rainfed	Irrigated	Rainfed
Wheat	–0.17	–0.18	–0.03	–0.11
Barley	n.a.	–0.10	n.a.	–0.12
Potato	–0.04	0.20	–0.02	0.05

Source: Based on DSSAT.

Note: Due to data constraints the DSSAT model was only run for the three above crops. In order to cover all the agricultural SAM sectors, yields for grains—other than wheat—followed barley yields and yields for all other crops followed yields for irrigated potatoes. n.a. = not applicable.

the economywide impacts of climate change on the Tunisian economy are negative. The estimates, however, vary significantly depending upon which GCM scenario is considered, and by global and local effects. It is also important to note that the model results should be interpreted as an optimistic scenario, in which the policy and economic environment allows for and supports climate change adaptation. Specifically, producers are assumed to be freely able to substitute labor, capital, land and inputs to react to changing relative costs of inputs, or imported food can replace locally grown food when relative prices of locally grown food increase (and vice versa). Under these assumptions, global climate change impacts translate to modest total losses in GDP of US$2–3 billion over the projection period of 30 years, comparing the economy to a situation of perfect mitigation (figure 3.4). Given that higher global food prices can benefit exporters of agricultural goods (and thus raise agricultural GDP), the combined effects of climate change will cost the Tunisian economy between US$2 and 2.6 billion. As Tunisia is also reliant on food imports, global food price increases outweigh the small benefits from higher export prices and so generate the greatest losses in the economy.

Figure 3.4 Economywide Losses in GDP

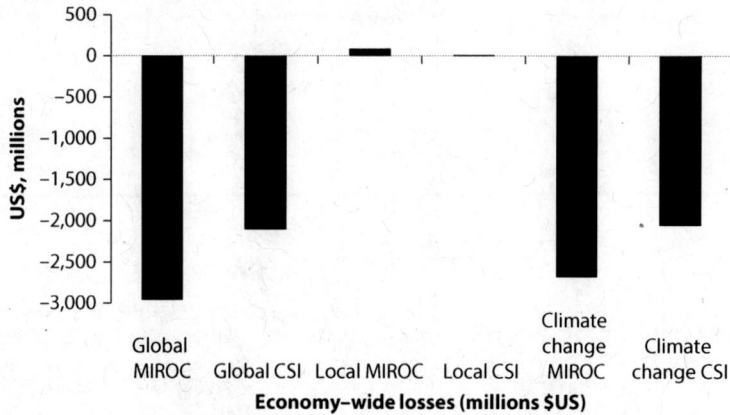

Source: Tunisia DCGE model.
Note: MIROC = Model for Interdisciplinary Research on Climate; CSI = Commonwealth Scientific and Industrial Research Organization.

Impact on Agriculture

Economywide losses are mainly driven by the climate change impacts on agriculture. Results of the DCGE model show that climate change-related global food price increases may have a slightly positive effect on the agricultural sector through higher returns to production factors. Agricultural activities benefit from the price increases, attract additional capital and labor, and thereby slightly increase production. The annual average agricultural growth rate is 0.1 percentage points higher compared to perfect mitigation across both GCM scenarios (figure 3.5). However, when local climate change is added on, these relatively small and price-induced positive effects are outweighed by the strongly negative effects of reduced yields where agricultural growth is projected to suffer reductions in yields under both the MIROC and the CSIRO scenarios. More specifically, results from the DCGE model show that agricultural growth may drop between 0.3 and 1.1 percentage points annually by the end of the study period. During the initial years, the losses are more severe, and agricultural growth recovers over time, that is the model contains some endogenous mechanisms for climate change adaptation. For example, people can freely adapt to a changing climate by switching crop patterns and moving out of agriculture and into other sectors of the economy that have development potential.

Combining local and global effects shows that the slightly positive effects of higher world market prices on agriculture do not cushion the large negative effects of lower yields enough. The overall effect of climate change on

Figure 3.5 Climate Change Impacts on Agricultural GDP

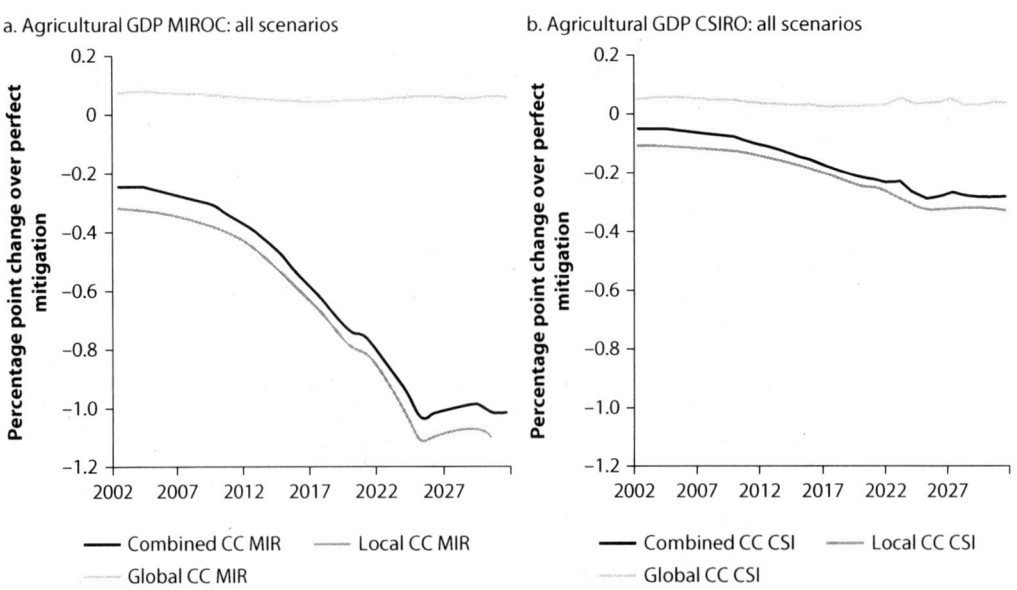

a. Agricultural GDP MIROC: all scenarios

b. Agricultural GDP CSIRO: all scenarios

Combined CC MIR — Local CC MIR
Global CC MIR

Combined CC CSI — Local CC CSI
Global CC CSI

Source: Tunisia DCGE Model.

Figure 3.6 Impacts of Combined Climate Changes on Household Incomes

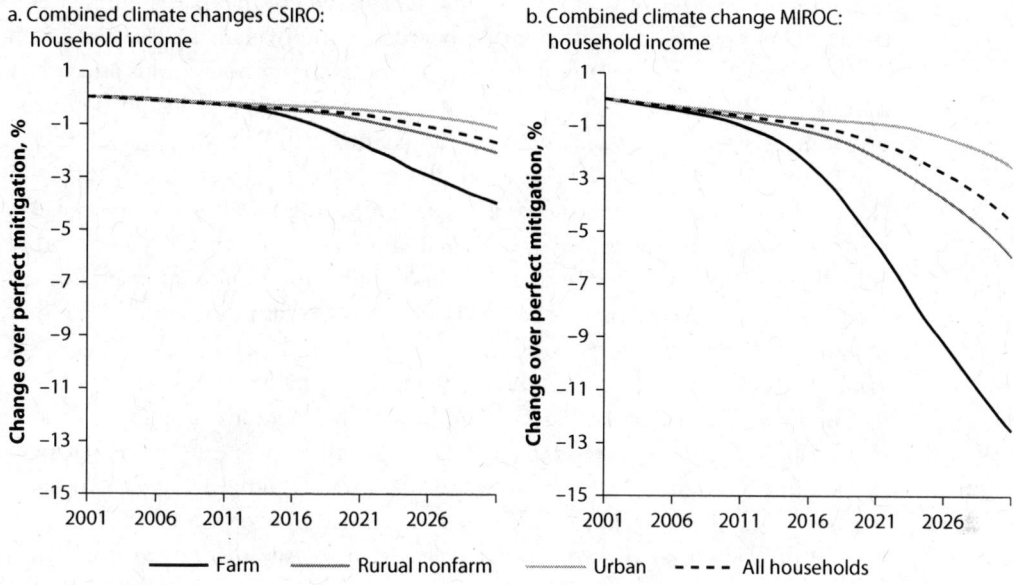

a. Combined climate changes CSIRO:
   household income

b. Combined climate change MIROC:
   household income

———— Farm   ———— Rurual nonfarm   ·········· Urban   - - - - All households

Source: Tunisia DCGE Model.

agriculture is negative, where despite the losses being slightly less than the losses to the sector under local climate change impacts alone, agricultural GDP still falls between 0.1 and 1 percentage points annually by 2030 for the CSIRO and MIROC scenarios, respectively. The overall impact on agricultural GDP due to global climate change impacts does not differ among the two GCM scenarios (figure 3.6). The difference appears when considering the local impact of climate change on the agricultural sector. The negative effects are amplified in that situation mainly due to the reduced crop yields.

Climate Change Impacts on Households

When analyzing the effects on households it is important to differentiate between rural and urban households and within rural households between farmers and other rural dwellers. In addition, when interpreting the results of the global and local scenarios (figure 3.6), it is important to keep in mind that climate change affects world food prices through changes in global production and consumption (global effect) and agricultural yields through changes in rainfall and temperature (local effects).

Under the global climate change scenarios, we find that farmers may benefit from higher food prices. However, the two scenarios give different results for farm households. Under the MIROC scenario, which projects higher international food prices than its CSIRO equivalent, the increase in the prices of agricultural products is welfare enhancing for those whose livelihoods depend on the agricultural sector, namely farm households. Under the CSIRO scenario, all households lose after several years of initial gains, where the increase in agricultural product prices is insufficient to overcome the overall increase in global food prices. The other households, namely rural nonfarm households and urban households, lose under higher world food prices as their real incomes decline due to higher food expenses under both scenarios.

The impact of lower yields (local climate change), however, is different (figure 3.6). Local climate change is welfare reducing for all household groups under both GCM scenarios; however, it is more so when we consider the MIROC scenario and when we also consider farm households. Households are affected through two major channels: (1) incomes of farm households see their incomes fall due to lower agricultural activity; and (2) lower yields raises domestic food prices, thus negatively affecting real incomes of households.

The long-term (local and global combined) implications of climate change in Tunisia lead to a total reduction of household incomes due to the stronger impact on household welfare through the local climate change impacts (figure 3.6). These welfare reductions accumulate over time. In 2030, all household incomes are projected to be 1–2 percent lower compared to a perfect mitigation scenario. Farmers are hardest hit. Negative effects under the MIROC scenario reduce farmer incomes by close to 13 percent by the end of the period compared to a scenario of perfect mitigation. However, under the CSIRO scenario, farmer welfare losses are less, reaching a more modest but still significant 4 percent reduction over perfect mitigation. Farmers suffer most from climate change in Tunisia, followed by rural nonfarm and urban households.

Notes

1. This section draws on Nelson et al. (2009).

2. The dynamic computable general equilibrium (DCGE) model described in the following section uses several outputs from the global partial equilibrium IMPACT model as drivers for agricultural and climate-change-related aspects. As a global-scale model, the climate change drivers in IMPACT are based on a resolution that is relatively coarse when compared with a medium-sized country such as the Republic of Yemen. Thus, even though the global projections are useful as the boundary conditions for the country-level CGE model, the production shifters for the intracountry regions can be improved upon, if sufficient local data are available.

3. Centre National de Recherches Météorologiques.

4. MIROC3.2 is the Model for Interdisciplinary Research on Climate developed at the Center for Climate System Research, University of Tokyo.

5. ECHAM5 model is the 5th generation of the ECHAM general circulation model developed at the Max Planck Institute for Meteorology, Germany.

6. In addition to various Global Climate Models (GCMs), Nelson et al. (2010) also include low, medium, and high assumptions on population and GDP per capita growth. For this study, we use the medium-level assumptions.

7. As previously described, variations in average monthly rainfall are compared with the equivalent baseline estimates.

8. Due to data constraints we could only do the DSSAT modeling for the three above crops. In order to cover all the agricultural SAM sectors, some assumptions about yields were made (table B.1).

References

Breisinger, C., and X. Diao. 2008. "Economic Transformation in Theory and Practice: What Are the Messages for Africa?" IFPRI Discussion Paper 797, International Food Policy Research Institute, Washington, DC.

Breisinger, C., O. Ecker, P. Al-Riffai, and B. Yu. 2012. "Beyond the Arab Awakening: Policies and Investments for Poverty Reduction and Food Security." IFPRI Food Policy Report 25, International Food Policy Research Institute, Washington, DC. (in English and Arabic).

Breisinger, C., T. Zhu, P. Al Riffai, G. Nelson, R. Robertson, J. Funes, and D. Verner. 2011. "Global and Local Economic Impacts of Climate Change in Syria and Options for Adaptation." IFPRI Discussion Paper 1091, International Food Policy Research Institute, Washington, DC.

Garnaut, R. 2008. "The Garnaut Climate Change Review." http://www.garnautreview.org.au/.

IPCC (Intergovernmental Panel on Climate Change). 2007. "Climate Change 2007— The Fourth Assessment Report of the Intergovernmental Panel on Climate Change." Cambridge University Press, Cambridge, UK.

Jones, J. W., G. Hoogenboom, C. H. Porter, K. J. Boote, W. D. Batchelor, L. A. Hunt, P. W. Wilkens, U. Singh, A. J. Gijsman, and J. T. Ritchie. 2003. "The DSSAT Cropping System Model." *European Journal of Agronomy* 18 (3–4): 235–65.

Jones, P. G., P. K. Thornton, and J. Heinke. 2009. "Generating Characteristic Daily Weather Data Using Downscaled Climate Model Data from the IPCC's Fourth Assessment." Unpublished Project Report. http://mahider.ilri.org/bitstream/handle/10568/2482/Jones-Thornton-Heinke-2009.pdf?sequence=3.

———. 2010. "Characteristically Generated Monthly Climate Data Using Downscaled Climate Model Data from the Fourth Assessment Report of the IPCC." CIAT. Unpublished. http://futureclim.info/.

Mougou, R., S. Rejeb, and F. Lebdi. 2002. "The Role of Tunisian Gender Issues in Water Resources Management and Irrigated Agriculture." The first Regional Conference on Perspectives on Water Cooperation: Challenges, Constraints and Opportunities. Workshop on Gender and Water Management in the Mediterranean, Cairo, Egypt, October 2002.

Nelson, G. C., M. W. Rosegrant, A. Palazzo, I. Gray, C. Ingersoll, R. Robertson, S. Tokgoz, T. Zhu, T. B. Sulser, C. Ringler, S. Msangi, and L. You. 2010. *Food Security, Farming, and Climate Change to 2050: Scenarios, Results, Policy Options.* Washington, DC: International Food Policy Research Institute. http://www.ifpri.org/sites/default/files/publications/climatemonograph_advance.pdf.

Nelson, G. C., M. W. Rosegrant, J. Koo, R. Robertson, T. Sulser, T. Zhu, C. Ringler, S. Msangi, A. Palazzo, M. Batka, M. Magalhaes, R. Valmonte-Santos, M. Ewing, and

D. Lee. 2009. "Climate Change: Impact on Agriculture and Costs of Adaptation." Food Policy Report, International Food Policy Research Institute, Washington, DC.

Rosegrant, M. W., S. Msangi, C. Ringler, T. B. Sulser, T. Zhu, and S. A. Cline. 2008. *International Model for Policy Analysis of Agricultural Commodities and Trade (IMPACT): Model Description.* Washington, DC: International Food Policy Research Institute. http://www.ifpri.org/themes/impact/impactwater.pdf.

Stern, Nicholas. 2007. *Stern Review on the Economics of Climate Change.* Cambridge, UK: Cambridge University Press.

Thurlow, J. 2004. "A Dynamic Computable General Equilibrium (CGE) Model for South Africa: Extending the Static IFPRI Model." Working Paper 1, Trade and Industrial Policies (TIPS), Pretoria, South Africa.

World Bank. 2007. "The Impact of Sea Level Rise on Developing Countries: A Comparative Analysis." Policy Research Working Paper WPS4136, World Bank, Washington, DC.

You, L., Z. Guo, J. Koo, W. Ojo, K. Sebastian, M. T. Tenorio, S. Wood, and U. Wood-Sichra. 2000. "Spatial Production Allocation Model (SPAM)." Version 3 Release 1. http://MapSPAM.info.

You, L., and S. Wood. 2006. "An Entropy Approach to Spatial Disaggregation of Agricultural Production." *Agricultural Systems* 90 (1–3): 329–47.

You, L., S. Wood, and U. Wood-Sichra. 2006. "Generating Global Crop Maps: From Census to Grid." Selected paper, IAAE (International Association of Agricultural Economists) Annual Conference, Gold Coast, Australia.

Yu, W., M. Alam, A. Hassan, A. S. Khan, A. C. Ruane, C. Rosenzweig, D. C. Major, and J. Thurlow. 2010. *Climate Change Risks and Food Security in Bangladesh.* London: EarthScan.

Yu, B., T. Zhu, C. Breisinger, and N. M. Hai. 2010. "Impacts of Climate Change on Agriculture and Policy Options for Adaptation: The Case of Vietnam." IFPRI Discussion Paper 1015, International Food Policy Research Institute, Washington, DC.

CHAPTER 4

Socioeconomic Effects of Climate Change in Central and Southern Tunisia

Photograph by Dorte Verner

Projected Climate Change in Central and Southern Tunisia and Impacts on Livelihoods

This chapter builds on step 1 of the Adaptation Pyramid related to assessing risks and opportunities by adding social analysis to the scientific and economic analysis provided in chapters 2 and 3. It examines the social implications of climate variability and change for the population of selected communities in seven governorates of central and southern Tunisia. It finds that food production systems and the agroecological conditions sustaining local livelihood strategies are not only severely stressed; the impacts vary socially and by area. The analysis shows that vulnerability is intrinsically related to the level of diversity of income sources available to the family, both within (in terms of diversity of products raised and sold) and outside agriculture. Vulnerability is also related to access to irrigated parcels and land ownership, which allows for a diversification of the water resource and production systems. The types and levels of vulnerability are related to adaptive capacity. The chapter also presents options for improving their resilience and adaptability to these phenomena recommending that local livelihood and production options should be diversified and local actors, governments, and sector institutions need support to implement identified actions. Uncertainty should be systematically integrated into planning, and investment in research, dissemination of information, awareness raising, skills training, and capacity building are all needed.

As with chapter 3 of this report, this chapter also focuses on an assessment of the risks and opportunities stemming from climate variability and change. This chapter, however, focuses on an analysis of the linkages between socioeconomic factors and vulnerability to climate change impacts, identifying the socially differentiated livelihood strategies, needs, and availability of assets that the rural population draws upon to respond to changing conditions. These conditions include natural resources affected by climate change impacts such as temperature variability, extreme events, long-term drought, and new precipitation patterns. The chapter begins by briefly introducing the climate change impacts on agricultural livelihoods projected for the central and southern regions, as well as the main socioeconomic characteristics of the population. The following section presents the methodological principles and data-gathering techniques used for the analysis in this chapter. The third section looks at how climate change has already affected regional livelihoods in pastoralism, oases production, arboriculture, and fisheries. The fourth section examines perceptions of and vulnerabilities to climate change impacts. The fifth section considers how certain livelihood strategies and assets can help populations adapt to climate change. The chapter concludes with operational recommendations for adaptive responses and institutional measures that can be applied in the context of the central and southern regions of the country.

These operational recommendations are also intended to assist with the prioritization of adaptation options in the Adaptation Pyramid (see figure 1.2 in chapter 1).

As shown in chapter 2, Tunisia is broadly divided into three main climate zones. A Mediterranean climate predominates in the region stretching along the northeast edge of the Atlas Mountains, while the central and eastern region of Tunisia is semi-arid, and the southwest and extreme southern region is arid. Impact studies that are region specific, such as this chapter, can provide vital input into locally relevant recommendations in terms of climate change adaptation strategies for vulnerable livelihoods and associated ecosystems.

The phase three study for the preparation of the Second National Communication to the United Nations Framework Convention on Climate Change (UNFCCC) from the Ministry of Environment (Ministère de l'Environnement et du Développement Durable/PNUD 2009) highlights that the central and southern regions of the country will be the most severely impacted by climate change.[1] For further details on climate change projections, we refer to chapter 2 of this report; in particular the sections on scenarios and potential impacts based on available climate data.

According to a preparatory study on the National Strategy for Adaptation of Tunisian Agriculture and Ecosystems to Climate Change from the Ministry of Agriculture (Ministère de l'Agriculture et des Ressources Hydrauliques/GTZ 2007), climate change would have direct impacts on agricultural production in the central and southern regions, particularly due to the projected increased occurrence of successive drought years.[2] These impacts could induce (1) an average 40 percent reduction by 2016[3] and 50 percent reduction by 2030–50[4] in olive oil production nationally; (2) an average 50 percent reduction (800,000 hectares) in the rainfed arboriculture surface area in the central and southern regions[5]; (3) an average 80 percent reduction in livestock (cattle, sheep, goats) in the central and southern regions by 2030[6]; and (4) an average 16 percent reduction by 2016 and 20 percent reduction by 2030 in the cereal production surface area in the central and southern regions, respectively.[7] Conversely, climate variability in terms of extremely humid spells could also bring some years of favorable rainfall.[8] However, the primarily declining trends in agricultural production projected for the central and southern regions will potentially negatively impact anticipated growth projections for these regions. Climate variability and change will ultimately determine the agricultural activities that will no longer be viable in economic terms.[9] Climate change impacts will also exacerbate already existing land and water resource management challenges in the central and southern regions. Box 4.1 illustrates some of the climate change impacts already felt in these regions, as recorded during field visits to the southern region in February 2010.

Box 4.1

Tunisia's Climate Change Hotspots: Arid Regions

Located in the arid regions of the country, the governorates of Kébili, Médenine, and Tozeur face particularly severe climate change challenges. Water scarcity and increases in extreme weather events are projected to have significant consequences for the agricultural sector, livelihood sustainability, and food security.

The impacts of water scarcity are already being felt in these governorates, which receive less than 100 millimeters of rainfall per year. Water resources are already mobilized to capacity, with 80 percent of oasis aquifers under extraction, coupled with some runoff harvesting (for instance, 6 million m3 mobilized out of 14 million cubic meters available in the Tozeur governorate). From 2002 to 2003, the Tozeur governorate witnessed annual rainfall amounting to less than 40–50 millimeters and a prolonged drought from 2002 to 2006. A drought in the Médenine governorate lasted from 1999 to 2002. Droughts can cause exacerbated aquifer and soil subsidence as well as salinization of soils in areas where water aquifers already have relatively high rates of salinity (up to 6–7 grams per liter).

Extreme weather events such as elevated temperatures, sandstorms, and sirocco winds are normal in the southern region. The Tozeur governorate, for example, regularly experiences 50–55°C temperatures during the summer, 74 days of sirocco per year, and 30–50 days of sandstorms. Climate change is projected to exacerbate the intensity and frequency of these events, with more consecutive hot days of 45–50°C, prolonged and consecutive droughts, and sudden precipitation. Rainfall in the Matmata area of the Médenine governorate has already noticeably changed, and in 2010 the area witnessed for the first time a sirocco wind and summer temperatures during the winter. The governorate also received 114 millimeters of rainfall in one hour during one winter at the end of the 1990s.

In the Tozeur governorate, oasis farmers must already balance irrigation availability for date palms and supplementing their revenue with other oasis crops. In the Douz region of the Kébili governorate, rangelands become degraded when native plants do not produce seeds as a natural mechanism to cope with drought.

Climate change impacts in these areas highlight the need for adaptive agricultural and pastoral practices and crops, and the diversification of agricultural and rural livelihoods. Additionally, adaptive measures such as soil and water conservation, replenishment of ground aquifer resources, and desalination methods need to continue to achieve sustainable extraction rates and protect resources.

Source: Field visits in the southern region, February 2010.

Climate change would negatively affect the already vulnerable livelihoods of rural populations in the interior central and southern regions Tunisia, as well as the agro-ecosystems on which they depend. These populations earn a significant portion of their income from agriculture, ranging from 13.7 percent of the active population in the Tataouine governorate to 30.4 percent in the Kasserine governorate, compared to a national average of 16.5 percent.[10] Unemployment rates are also significant, ranging from 12 percent in the Kébili governorate to 21 percent in the Gafsa governorate, compared to a national average of 16.4 percent. Rates are particularly high among women, ranging from 19.7 percent in the Gabès governorate to 28.3 percent in the Gafsa governorate.[11] The population in these regions is young, as in the rest of the country, with over 60 percent under 30 years of age.[12] Recently, the need for access to basic infrastructure and employment drove the local population to actively participate in fomenting the Tunisian Revolution in the governorates of Sidi Bouzid and Kasserine, considered the birthplace of the movement. The population's inability to cope with compounded hardships could accelerate the already-significant rural-to-urban migration phenomenon, which could weaken social and community structures. Box 4.2 provides key demographic data for the governorates visited during the data collection phase for this chapter, according to the National Census of 2004.

Rural populations in central and southern Tunisia are especially vulnerable to the impacts of climate change compared with their urban or coastal counterparts. Important work and research has been completed and is currently underway at the national level addressing the vulnerability of coastal areas,[13] urban areas,[14] and the northern region[15] to climate change impacts. This chapter thus

---

Box 4.2

Key Socioeconomic Characteristics for Governorates Visited in the Central and Southern Regions

a. Active Population (>15 Years of Age) Working in Agriculture, Forestry, and Fisheries

Governorate	Number	Percentage of governorate's active population
Gabès	15,203	17.1
Gafsa	10,696	15.0
Kasserine	28,485	30.4
Kébili	59,617	16.4
Tataouine	3,949	13.7
Tozeur	5,387	20.5

Source: National Statistics Institute 2004 Census.
Note: The terms agriculture, forestry, and fisheries include the exploitation of animal and vegetation-based natural resources, encompassing such activities as cultivation, animal raising, gathering of wood and other plants, and production of animals or animal-based products on-farm or in their natural habitat. At the national level, 16.5 percent of the population in engaged in these activities.

(box continues on next page)

Box 4.2 Key Socioeconomic Characteristics for Governorates Visited in the Central and Southern Regions *(continued)*

b. Unemployment Rate (>15 Years of Age)

Governorate	Men	Women	Governorate average
Gabès	13.7	19.7	15.1
Gafsa	19	28.3	21.1
Kasserine	18.1	25.3	19.8
Kébili	9.3	22.2	12.0
Tataouine	13.4	23.4	15.3
Tozeur	13.1	25.7	16.0

Source: National Statistics Institute 2004 Census.
Note: The national unemployment rate is 16.4 percent.

c. Age Structure of the Population (Under 30 Years of Age)

Governorate	Percentage under 15 years of age	Percentage between 15 and 29 years of age
Gabès	27.6	30.6
Gafsa	28.5	30.4
Kasserine	33.0	29.7
Kébili	27.8	32.3
Tataouine	31.7	30.9
Tozeur	28.3	29.1
National average	26.8	29.5

Source: National Statistics Institute 2004 Census.

complements this work in focusing on the central and southern regions. Governorates of the central region include Gafsa, Kasserine, Kairouan, Mahdia, Monastir, Sfax, Sidi Bouzid, and Sousse. Governorates of the southern region governorates include Gabès, Kébili, Médenine, Tataouine, and Tozeur (see map on page i).

Methodological Principles and Data-Gathering Techniques

This chapter illustrates the dynamics of climate change impacts and the related dynamic adaptation strategies of rural households. It is based on an examination of change over time, by comparing "how it is" with "how it was." To this effect, the research established three reference points on a timeline, and attempted to illustrate change in these dynamics over a span of 25 years. The timeline included One Generation Ago (1991), Now (2011) and Near Future (2016).

The years were chosen to reflect change over a generation (1991–2011) and envisage prospects for the near future (2011–16). As informants cannot be expected to recall a specific year, culturally appropriate ways of accessing the information were used (that is, how was it when your first child was born, how do you think it will be when your son is a grown man). The research does not aim to acquire information about a specific year but rather about a "notion of past." Information about the past was further captured with reference to story-telling or events.

The field research comprised a series of diverse information-gathering approaches that provided insights into causalities, connections, trends, and patterns. For instance, household surveys used a questionnaire format combining open-ended with closed questions to capture qualitative and quantitative data (appendix E). Limitations in field time and resources precluded attempts to produce statistically significant findings. Rather, the data gathered are intended to enrich qualitative information with graphic and quantitative illustrations of local perceptions.

The analysis considers communities that vary in size and face differing socio-economic, agroecological, and climate conditions, so that the socially and culturally differentiated use of and preferences for various adaptation strategies may be fully understood. To ensure that all relevant social groups were considered, a maximum variation sampling strategy was employed that covers the relevant aspects of adaptation to climate change impacts. Multistakeholder and regional level government technical staff workshops of three to four hours were combined with one-hour interviews with individuals with different perspectives. A "snowballing until saturation" selection of the latter interviewees was used, asking guides and informants to point to other individuals/households, who respond differently to the locally perceived climate change impacts, and then making sure these informants are included in the survey.

The study aimed for representation of both men and women in all categories of well-being. The selection of subjects was intentionally slightly biased toward older respondents, as they have historical experience and are traditional holders of knowledge who could provide details about changes over time.

The technical departments (livestock, agricultural production, water management, and so on) of the Regional Commissions for Agricultural Development (CRDAs)[16] in each of the seven governorates visited were consulted during seven workshops lasting three to four hours. CRDAs are the regional representations of the Ministry of Agriculture (one in each governorate) and are one of the first points of contact for local rural populations, particularly in terms of agricultural extension services, and other programs. The opportunity was used to inquire into vulnerabilities, institutional constraints, and opportunities regarding climate change impacts as well as general use of and pressure on natural resources. A realistic climate change impact scenario for each governorate was used based on the draft upcoming National Climate Change Strategy and preliminary analysis for chapter 2.

Box 4.3

Selection Criteria for Interviewees and List of Surveyed Communities

Number	Region/ governorate	Community/ site name	Region[a]	Agroecological zone[b]	Main productions/ sources of income[c]
1	Kasserine	Fej Bouhsine	CR	SA	AP, RA
2	Kasserine	Younes El Kahari	CR	SA	AP, RA
3	Kasserine	Feriana	CR	SA	AP, RA
4	Gafsa	El Guettar	SR	A, O	AP, RA, IA, O
5	Gafsa	Lalla	SR	A, O	AP, RA, IA, O
6	Gafsa	Oasis	SR	A, O	AP, RA, IA, O
7	Tozeur	Maksem Echik	SR	A, O	AP, RA, IA, O
8	Tozeur	Hazoua	SR	A, O	AP, RA, IA, O
9	Tozeur	Old Oasis	SR	A, O	AP, RA, IA, O
10	Kébili	GDA Douz Oasis	SR	A, O	AP, RA, IA, O
11	Kébili	Douz Centre	SR	A, O	AP, RA, IA, O
12	Kébili	Zaafrane Douz	SR	A, O	AP, RA, IA, O
13	Medenine	Dergoulia	SR	A	AP, RA, IA
14	Tataouine	Tlelett	SR	A	AP, RA
15	Tataouine	Ksar Ouled Dabab	SR	A	AP, RA
16	Tataouine	Kasbet Ajlat	SR	A	AP, RA
17	Médenine	DJerba -Houmet Souk	CC	SA	AP, RA, IA, O, F
18	Médenine	Robbana-Thelit DJerba-Midoun	CC	SA	AP, RA, IA, O, F
19	Médenine	Houmet Souk Port	CC	SA	AP, RA, IA, O, F
20	Gabes	CRDA	CC	SA	AP, RA, IA, O, F

Source: November 2011 field visit.

a. Central Region (CR), Southern Region (SR), Central Coast (CC).

b. Arid (A), Semi-arid (SA), Access to Oasis (O).

c. Agropastoralist (AP), Rainfed (RA) and Irrigated (IA) agriculture, orchards (dates, figs, olives etc.) (O), Fisheries (F).

A total of 51 household interviews were held (20 group interviews and 31 individual interviews). Group interviews often involved members of Agricultural Development Groups (GDA),[17] which are community-based organizations that often represent the first point of contact for the CRDA with the rural population for agricultural extension services. Box 4.3 includes the name and governorate of the communities visited, as well as their key social, economic, and agroecological characteristics. Box 4.4 provides anecdotes from the interviews with both community groups and individuals with differing vulnerability types.

The climate change impact scenario was related to a high level of specificity with regard to relevant aspects and themes. However, interviews were held in an open-ended fashion, to avoid introducing biases or eliciting responses designed to please the interviewer.

Box 4.4

Surveying Communities in the Central and Southern Regions

Individual Respondent in Gafsa Oasis, Gafsa Governorate

Source: Jakob Kronik
Note: Photo 2 shows an example of the " least vulnerable" category of
respondents due to diversified agricultural production (date plams, fruit trees,
vegetable and fodder crops, livestock), land ownership, and relatively easy
access to credit.

**Group Interview Respondents in the
Dergoulia Community, Médenine
Governorate**

Source: Jakob Kronik

**Individual respondent in
El Guettar community, Gafsa
governorate**

Source: Jakob Kronik
Note: Example of a "medium vulnerability"
category of respondents with some
diversification in revenue (pension, olive trees,
and date palms). However, children are
unemployed, hence a lack of fixed salary
within the family.

(box continues on next page)

Box 4.4 Surveying Communities in the Central and Southern Regions. *(continued)*

Individual respondents in Zaafrae community, Kébili governorate (right)

Sorce: Jakob Kronik

Note: Example of a "most vulnerable" category of respondents with sole source of revenue (date palms) due to water scarcity, where there was diversity before (vegetable and fodder crops), and difficult access to credit due to lack of land ownership.)

The scenario discussion was introduced based on reflections of present and past, and of the scenarios of potential future climate change impacts. Questions raised included the following:

- What do people do now when they experience this climate change impact?
- What did people do in the past when they experienced this climate change impact?
- Which resources (assets) do/did they draw upon?"
- Which assets are/were temporarily/finitely damaged?
- Which resources (assets) can/could they substitute with?
- What do you think would have improved your ability to cope and adapt during that climate event?
- What will happen to the ways you live if droughts become longer, rains less predictable but more intense, and floods more common?

During both the household interviews and the workshops, issues about the institutions and networks that some individuals and families are able to draw upon to increase the effect of their coping strategies were addressed, and the answers were related to the stakeholder/institutional mapping.

Appendix E provides a more detailed explanation of the methodology used for the analysis in this chapter as well the questionnaire used for household interviews.

Climate Change Impacts on Livelihoods

Impacts on Agro-Pastoral Livelihoods

Collectively, central and southern Tunisia is home to much of the country's livestock: 24 percent of national cattle stocks, 62 percent of national sheep stocks, and 76 percent of national goat stocks.[18] The central governorates of Sid Bouzid, Kairouan, Kasserine, and Sfax together account for 34.1 percent of the national sheep stock, while the southern governorates of Gafsa, Médenine, Gabès, and Tataouine account for 17.6 percent.[19] The southern region accounts for 55.5 percent of national goat stocks, with 42.8 percent raised in the Médenine, Tataouine, and Gabès governorates alone.[20] Some rural populations depend entirely on pastoral activities for their livelihoods, thus mitigating rural outmigration.[21] However, climate change impacts and human pressures on local ecosystems can increase the vulnerability of these pastoral communities, by affecting their capital stocks, hindering coping strategies, decreasing the productive performance of livestock, and generating tensions with other herder and farming communities.[22]

Natural rangeland degradation in the central and southern interior regions is mainly the result of decreasing and more-variable rainfall, coupled with the eradication of natural vegetation and subsequent erosion. The expansion of crops and arboriculture, overgrazing, and a detrimental use of machinery all play a role in land degradation.[23] Overgrazing rates in central and southern Tunisia are estimated at 78 percent and 80 percent, respectively.[24] These rates have doubled in the past 25 years in the central region and in the past 40 years in the southern region.[25] Twenty percent of rangelands have already disappeared. Pastoral functions in steppe ecosystems are projected to significantly decrease in the central region and disappear altogether in the south by 203050.[26]

In the Kasserine governorate, for example, 159,000 hectares of alpha grass rangelands currently cover the southern areas of the governorate, including Kasserine South, Sbeitla, Feriana, and Chaambi National Park.[27] Local officials believe that the quality and quantity of the vegetation cover of natural rangelands were much higher 20–30 years ago, due to a combination of more abundant and regular rainfall during wet years and less destructive anthropogenic pressures.[28] In the Gabès governorate, animal needs have surpassed the capacity of the rangelands to produce for the past 30 years or so.[29] In the Gafsa governorate, 400,000 hectares out of 780,000 hectares of rangelands are considered degraded, with 100,000 hectares at an advanced stage of degradation.[30] Natural rangeland degradation is also further exacerbated by the increasing presence of invasive species. Box 4.5 describes some of the impacts of native species loss on natural rangelands.

Overgrazing in semi-arid and arid ecosystems increases soil vulnerability to erosion by wind during droughts, and by water during storm events.[31] Current estimates show that about 50 percent of the country's arable lands are already eroded.[32] Sand encroachment due to climate change and land degradation is also a major issue. In the Médenine governorate, Ben Gardane and Médenine have witnessed this phenomenon since the 1980s, and 50 percent of the lands in the governorate is now considered threatened.[33] Extreme events induced by climate change could accelerate the soil degradation cycle, resulting in a

Box 4.5

Loss of Native Species: A Major Factor in Rangeland Degradation

A major factor in natural rangeland degradation is the loss of native species, through a combination of overgrazing coupled with irregular rains and temperatures.[a] In Tunisia's arid areas, native plants cope with drought by foregoing seed production.[b] Prolonged drought thus results in an ever-diminishing resource.

During years of normal rainfall, rangelands adjacent to oases fare well due to the microclimate within and around the oases.[c] With irregular rainfall and temperatures, however, rangelands have become degraded and annual plants have disappeared altogether, affecting soil quality.[d] When the rain does come, it is torrential over the course of a month or so.

Consequently, the vegetation cycle does not last long and does not have an opportunity to regenerate itself.[e] As native species become scarcer, rangelands become more susceptible to invasive species, a phenomenon that has affected rangelands across the central and southern regions. In the Menzar Habib area of the Gabès governorate, invasive species that are unpalatable to livestock have been a growing concern over the past 40 years.[f] There is a need for a better understanding between the phenomenon of invasive species and climate change, as well as an emphasis on monitoring of native species health.[g]

Degraded Rangeland in the Southern Region

Source: Viviane Clément.

Source: Field visits to central and southern Tunisia, February 2010.

a. From interview with the Gafsa CRDA in November 2011.
b. Notes from field visit to Kébili in February 2010.
c. From interview with Agricultural Development Group (GDA) in Hazoua, Tozeur governorate in November 2011.
d. From interview with Agricultural Development Group (GDA) in Hazoua, Tozeur governorate, and with the Médenine CRDA in November 2011.
e. Ibid.
f. From interview with the Gabès CRDA in November 2011.
g. Ibid.

decrease in both revenues and yields, and prompting an extension of cultivated lands and soil saturation.[34] Additionally, the pursuit of marginal and episodic cereal cultures could further expose central and southern steppe ecosystems to erosion.[35]

The degradation of natural rangelands, exacerbated by climate change impacts and the disappearance of key rangeland ecosystem functions, could drive local residents to completely rely on outside feed for their animals and/ or migrate to more northern pastures. This could result in further overgrazing despite current rehabilitation and conservation programs.[36] In the past, agro-pastoralists who grew crops and raised livestock could rely on natural range-lands to feed their animals, and fodder reserves could be built up during wet years in anticipation of less-abundant dry years.[37] At present, with less rain-fall, increased climate variability, and shrinking rangeland areas, current live-stock populations in the central and southern governorates cannot be sustained without government-subsidized supplementary feed and concen-trates.[38] This means an increase in the production cost of animals. This is particularly detrimental to small agro-pastoralists, who are unable to build sufficient feed reserves and are caught in a vicious circle where fodder (par-ticularly barley) produced during wet years is immediately sold, only to be repurchased as supplementary feed for a higher price during dry years.[39] Given four to five consecutive dry years, animals must be sold at reduced prices due to a lack of resources to purchase inputs.[40] Agro-pastoralists who move their livestock from one grazing area to another on a seasonal basis (transhumance) must go ever farther in search of pasture. Some travel from the central and southern governorates all the way to Tunisia's northwestern region.[41] Generally, with increases in production costs, small agro-pastoral-ists are finding it more difficult to maintain their animals and face ever-decreasing production.[42] Increased production costs also imply increases in the price of meat, an additional burden on poorer rural households.[43] In certain governorates such as Tataouine, traditional pastoralists have had to transition to agro-pastoralism, and are currently moving more toward irri-gated agriculture only.[44]

Impacts on Oasis Livelihoods

Oases are unique agroecosystems that epitomize human adaptation to arid areas through the optimization of scarce natural resources. Oases cover 57 percent[45] of the Kébili governorate's total area, which hosts 56 percent of the date palms in Tunisia.[46] The governorate is responsible for 70 percent[47] of the national production of the *deglet noor* variety of dates and 56 percent of the national date production, generally.[48] The traditional three-tier cul-tivation system of the central and southern oases, including the Gabès,

Gafsa, Jérid, and Nefzaoua oases, has enabled an integrated production of dates (top tier), fruit trees (middle tier), and vegetable and fodder crops (bottom tier) through efficient small-hydraulics systems and the maintenance of a favorable microclimate within the oases. As further illustrated in box 4.6, this traditional cultivation system contributes to Tunisia's food security by yielding diverse crops destined for both export and local consumption.[49] Domesticated animals such as sheep, goats, donkeys, mules, horses, rabbits, and chickens also contribute to both a balanced ecosystem and self-sufficiency.[50]

The oases hold a rich store of human cultural and ecological knowledge. Water resources are often managed through traditional communal irrigation systems, based on water distribution between parcels using a complex network of channels.[51] Oases have been central to the development of the areas in which they are situated, providing employment opportunities and opening the door to a number of related activities, including commerce in fresh and dried fruits, vegetables, cereals, alfalfa, dried forage, seeds, and fertilizers; handicrafts such as basketry and wickerwork; and traditional blacksmithing and carpentry.[52]

Oases in central and southern Tunisia are particularly vulnerable to degradation stemming from water scarcity and encroachment by urban areas and sand dunes. These pressures will only be exacerbated by climate change.[53] This is the case for the Gafsa Oasis and Gabès Oasis, one of the last coastal oases (see box 4.6 for more details).[54] In the Tozeur governorate, traditional three-tier oases thrived up to the 1970s and 1980s.[55] With increasingly irregular and generally declining rainfall and the progressive expansion of the cultivated oasis area, the "tour d'eau" or water turn to each individual parcel within the oases is ever lengthening, and the three-tier system is disappearing, as producers focus resources on *deglet nour* date palm production.[56] Vegetable and fodder crops, then fruit trees are the first to go, as these die out or become diseased.[57]

Water scarcity in the oases can in turn lead to soil degradation through salinization, since soils are not flushed out, and affect yields.[58] However, soil salinization can also occur through the evaporation of excess irrigation water, often associated with inefficient use and poor soil drainage, which leaves dissolved salts in the soil.[59] In the Tozeur area, an estimated 68 percent of irrigation water fulfills crop needs, while 18 percent drains out.[60]

Sand dune encroachment into oases, a recurrent and normal phenomenon, can be exacerbated by the degradation of the vegetation cover in surrounding rangelands, which results from prolonged drought or overgrazing. When this happens, sandy oasis soils are exposed to erosion and further degradation.[61] Mitigation measures for such encroachment over the past 50 years have included artificial protective dunes, dune fixation, and windbreakers.[62]

Box 4.6

Gabès Oasis: One of the Last Coastal Oases

The Gabès oasis is one of the last of its kind in the world and currently undergoing the application process to become a UNESCO World Heritage site. The climate of the 7,000 hectare oasis is influenced by the Sahara Desert and the Mediterranean Sea. It has three tiers of crops: the top tier consists of date palms (45 varieties have been recorded), the middle tier is fruit trees (including pomegranate, apricots, figs, apples, grapes, peaches, and mulberries), and the bottom tier horticultural plants (carrots, turnips, onions, tobacco, alfalfa, and *henna*, which is typical of the Gabès area). This integrated vertical structure constitutes a very favorable microclimate for biodiversity, representing a refuge for small mammals, reptiles, mollusks, and insects as well as an invaluable rest stop for trans-Saharan migrating birds. Additional research could provide a more accurate valuation of biodiversity and associated ecosystem services within the oasis.

However, this diverse agroecosystem is facing a number of challenges on several fronts, primarily water scarcity, soil salinization, and urban encroachment. Water springs and surface groundwater aquifers have now been exhausted, and the oasis is currently supplied by groundwater extraction from three deep wells. The intrusion of saline seawater into deep coastal aquifers, resulting from a combination of sea-level rise due to climate change and an

Degraded Three-tier System in the Gabès Oasis

Source: Viviane Clément.

(box continues on next page)

Box 4.6 Gabès Oasis: One of the Last Coastal Oases *(continued)*

overexploitation of groundwater aquifers, has led to decreased water quality, and soil salinization. Demand for groundwater resources is high, originating from the oasis itself, other agricultural activities, and the industries in the area, including chemical (phosphoric acid and phosphate) and cement industries. Further, more urban dwellings are being constructed within the oasis itself, and producers now tend to eliminate date palms in favor of horticultural crops, as the value of dates produced in the oasis has declined due to competition with the more prized *deglet noor* dates grown in more southerly regions. The loss of date palms means a loss of important regulating ecosystem services for the preservation of the microclimate and protection from dry winds. These challenges will only be exacerbated by climate change.

Potential adaptation options centered on integrated water resource management could include (1) improved methods for water harvesting (currently, only 85 percent of these waters is collected); (2) groundwater aquifer recharge using water runoff with input from the Institute for Arid Regions /*Institut des Régions Arides* for recharging; (3) increased efficiency in water use by the industrial and tourism sectors; and (4) treated wastewater reuse options in collaboration with a National Office for Water Treatment/*Office National de l'Assainissement* (desalination). A management plan for the oasis is being prepared through the Gulf of Gabès Marine and Coastal Resources Protection Project supported by the GEF and managed by the World Bank.

Source: Field visits to the area in February 2010; UNESCO 2010.

Effects on Arboriculture Production

Tunisia's olive tree plantations occupy a third of all arable land in the country, at around 1.6 million hectares.[63] The central and southern regions account for 70 percent and 18 percent of the national olive arboriculture area, respectively.[64] Olive tree arboriculture represents an ancient and traditional livelihood in Tunisia, at the forefront of economic and social life. Indeed, 57 percent of the country's producers derive all or most of their revenue from this activity.[65] Olive oil makes up 50 percent of food-related exports and the domestic market for olive oil is between 50,000 and 70,000 tons per year.[66] Box 4.7 provides additional information on climate change impacts on olive oil production for a small producer.

Olive arboriculture plays a crucial role in regional stabilization, as it often represents the only viable agricultural activity in certain areas where water is scarce and soil is poor. In this way, it mitigates rural outmigration while contributing to erosion prevention through vegetation cover.[67] However, olive arboriculture depends greatly on inter- and intra-annual rainfall availability and is thus susceptible to drought. For instance, national production fell to 255,867 thousand tons and 205,867 thousand tons in 1995 and 2002, respectively, due to drought (down from an average of 555,867 thousand tons per year).[68]

Box 4.7

Climate Change Impacts on Olive Arboriculture: Changing Production Cycles

An individual interview with a small producer in Djerba, Médenine governorate, gave additional insight on the impacts of climate variability and change on olive oil production. During dry years, oil production decreases and production cost increases. Fruits are smaller and "drier" and thus added pressurization is needed in the extraction process, adding to production costs. Furthermore, as olive growers often demand oil extracted from the fruits of their specific orchard, dry years make it more difficult to obtain a critical mass of olives to transform from a given grower, also adding to production costs. Issues of soil health and water quality also affect both the yield and quality of the olives and thus on olive oil. In these conditions, most of the revenue gained from production goes toward running costs with limited profit. Olive oil producers are intrinsically linked to olive growers through the quantity, quality and price of extracted olive oil.

Temperature variability has also impacted olive oil production. The flowering of olive trees in 2011 was observed in November (instead of March 2012, so early flowering). This, coupled with the lack of rainfall, will shorten the harvesting campaign, from November to December instead of November to January. Additionally, new diseases have been observed during times of drought, leading to a reliance on chemical pesticides.

If such trends continue, producers will have to sell their product at lower-than-production cost value to repay loans taken over the course of the year. Climate change–related challenges add to the already-existing pressures of competition on international markets, and a general lack of quality standards.

Olive trees flowering in November instead of March in Djerba, Médenine governorate

Source: Jakob Kronik.

Source: Interview with olive oil producer in Djerba, Médenine governorate, in November 2011.

Olive tree production in the central and southern regions is likely to be affected by drought and water scarcity induced by climate change. Vulnerabilities as well as adaptation strategies will depend on the type and age of individual trees, as well as the methods used to irrigate them.[69] In the Matmata area of the Gabès governorate, reductions in rainfall have prompted the CRDA to persuade farmers to trim down their olive trees (200,000 trees) to reduce production costs and ensure the trees' survival, despite existing soil and water conservation works.[70] In the Gafsa and other governorates, punctual irrigation interventions have been necessary to save olive tree plantations from drought conditions.[71] In the Tataouine governorate, the last six years of harvest have not been favorable, and olive oil stocks are practically depleted.[72] In the Gabès governorate, production has never been so reduced as in recent years, and supplemental irrigation has been needed more frequently, particularly for more marginal areas.[73] In the Médenine governorate, olive tree plantations were in critical condition during the summer of 2011, and even with ensuing autumn rains, have not all recovered.[74] For the Médenine governorate's four million trees,[75] as for plantations in other governorates, the main challenge is the development of adaptation measures to cope with the new irregular cycles in rainfall.[76]

Other types of arboriculture are also likely to be affected by climate change impacts, especially rainfed species such as almond and fig trees, which are less drought resistant in semi-arid and arid areas.[77] In the Kasserine governorate, projections to 2030 show a 50 percent decrease in arboriculture production, including that of native species.[78] The central and southern interior regions are finding rainfed arboriculture, and rainfed agriculture more generally, to be increasingly unprofitable.[79] In Djerba, farmers who used to plant wheat and sorghum during rainy years now obtain comparable yields for this production only one out of every five years.[80] In the Kasserine governorate, a 1999–2001 drought depleted seedling stocks,[81] and projected climate change impacts for the governorate to 2030 show a 20 percent decrease in cereal production.[82] This was also the case with the Tataouine governorate, where local seed stocks have been lost, and no cereals have been cultivated in the last four to five years of continuing drought.[83] The central and southern governorates are looking to expand irrigated perimeters for agriculture and arboriculture, although this raises sustainability issues with regard to efficient water use and declining groundwater resources.

Impacts on Fishery Livelihoods

Coastal fisheries are a particularly important mainstay livelihood around the rich coastal areas of the central and southern regions. The national fishery fleet consists of around 11,000 fishing units, including around 400 trawlers.[84] Half of all fishing units operate in the Gulf of Gabès, one of the nation's most important fisheries.[85] Most fishermen along the Gulf of Gabès coastline are small coastal fishermen (76 percent).[86] They compete vigorously with large trawler and sardine boats based out of the Gulf's main port cities, especially as fish stocks continue to decline from overfishing, the flouting of regulated fishing zones, and use of invasive fishing methods.[87] In the Gulf, stocks are already overfished by 30 percent and biodiversity is declining, paralleling reductions in submerged aquatic vegetation cover resulting from invasive fishing methods, particularly trawling.[88] In the Djerba port of Houmet Souk, the fishing high season from September to November in 2011 brought in 120 tons, compared to the usual 150 ton average.[89]

Invasive species represent an added pressure on fisheries, migrating mainly from the Red Sea through the Suez Canal.[90] One example in the Gulf of Gabès is the *crevette blanche* (white shrimp), which is now found all over the Mediterranean.[91] The species outcompetes the native *crevette royale* (royal shrimp), which has more commercial value for fishermen.[92] Another invasive species is the Nile tilapia, which died out in the 1960s but returned in the 1980s. It is now found in the drainage of oases waters, or in *oueds* (seasonal streams) that communicate with the sea.[93] The Nile tilapia's impact on local ecosystems remains to be further explored.

The direct impacts of climate change on fisheries are not yet well understood. However, it is clear that ecosystem changes due to temperature increases or the submersion of wetland ecosystems, which act as natural nurseries, may affect marine biodiversity and lead to a decline in traditional fishery species along coastal areas.[94] An important phenomenon associated with increased water temperatures is algal blooming, which now consistently occurs in areas where water is relatively shallow and circulation is slow (for instance, in the Boughrara Lagoon off Djerba).[95] Other phenomena that may have linkages to climate change include (1) an increasingly late start to the sardine fishing season, as the species lays its eggs near the coast and is possibly influenced by fluctuating temperatures; and (2) an association between rainfall and the organic/mineral content of coastal waters, which affects biodiversity health.[96] Box 4.8 illustrates the climate change- and management-related challenges to fishery livelihoods in two important wetland ecosystems in the Gulf of Gabès: the Bougrara and Bibans lagoons.

Box 4.8

Fisheries in the Gulf of Gabès and Wetland Ecosystems

The Gulf of Gabès is one of Tunisia's most important areas for fisheries production. The Gulf accounts for 40 percent of national production and remains a crucial nursery ecosystem for the Mediterranean. It yields around 8,000 tons of fish per year on average with the bulk of activity from sardine fishing. Other production seasons include blue fish in the summer. However, this resource is facing a number of challenges stemming from overfishing and the declining quality of marine waters, from industrial pollution from phosphate treatment plants, food industries, and tanneries (industrial zones in Sfax, Skhira, Gabès, and Zarzis to a lesser extent); tourism (Djerba and Zarzis), port activity (three commercial ports, one petroleum terminal, three offshore fishing ports, one coastal fishing port and three fishing shelters); and urban areas.

(box continues on next page)

Box 4.8   Fisheries in the Gulf of Gabès and Wetland Ecosystems *(continued)*

The Gulf encompasses two key lagoon ecosystems, the Bougrara and Bibans lagoons. The Bougrara Lagoon is the largest in Tunisia, covering around 50,000 hectares, and is home to 72 species of fauna, including some 30 fish species. Seventy-eight percent of Tunisia's lagoon fishermen are based here. Pressures on the lagoon include pollution from two large aquaculture farms, a desalination plant, and a slaughterhouse. The Bibans Lagoon covers around 30,000 hectares and is under pressure from overfishing, as well as pollution pressures from domestic waste and a slaughterhouse. These ecosystems are crucial from a biodiversity standpoint, serving as nurseries for a number of fish species.

In the Bibans Lagoon, local fishermen have devised a sustainable fishery with a weir catch system which consists of a system of fences and cages that trap fish on their way out from the lagoon into the Gulf. The fishery has also established no-fishing periods to ensure replenishment of stocks. However, this system is under pressure from unsustainable fishing methods and levels further out into the Gulf, which jeopardize the fishing stocks coming back into the lagoon. From a social perspective, the fishing population is aging, with young people migrating to Libya. Main climate change impacts on the fisheries of the Gulf of Gabès include the submersion and subsequent restructuring of humid ecosystem zones, including the Bibans and Bougrara lagoons and the low-lying Kerkennah and Djerba islands, as well as coastal erosion, which endangers key infrastructure.

Source: Notes from field visits to the area in February 2010; Ministère de l'Environnement et du Développement Durable, 2008.

Perceptions of and Vulnerabilities to Climate Change Impacts

Climate change is mentioned and perceived, although differently, in most communities visited in Tunisia. The overall perception is that the climate is not behaving as it did in the past: there is less rainfall, an increase in temperature, an increase in climate variability, increasingly erratic onsets of the rainy season, more frequent sandy winds (generally associated with drought), and a general feeling that droughts occur more frequently nowadays than the past. Meetings and interviews conducted with local population and local authorities have made clear, that for the past two decades, drought has not been regarded as a temporary climatic phenomenon, but instead as a structural factor that should be integrated into development strategies for the central and southern regions. Degrees of changes in weather conditions and rainfall patterns are recalled in terms of growing variability and instability in agricultural production, diminished water flow in *oueds*, and drying of fig and olive trees. Science confirms these community observations to an extent, as the Tunisian climate has shifted towards longer dry periods (sequences of years), although within an overall historical trend of oscillation.

Semi-Arid Areas of the Central and Southern Governorates

In the semi-arid areas of the central and southern governorates, rainfall usually comes in the spring and fall, with January to March and October to December being the wettest months.[97] During the rainy season, storms can bring significant amounts of rain within just a few hours.[98] The driest months are April through September, with peaks of dryness during July.[99] In these semi-arid areas, drought is a recurrent phenomenon, with one to two wet years interspersed between three to four dry ones. However, there is a general sense that seasons have been altered, affecting the natural cycles in the region's agroecosystems.

Traditionally, agro-pastoralists would build fodder reserves from the natural rangeland during wet years to sustain their herds through the dry years. However, problems arise when several periods of drought become consecutive, without sufficient rainfall in between. During such years, available grazing areas become scarcer, and more resources are spent on supplemental feed and concentrates. Herders must often sell off heads of livestock at reduced prices to raise funds for the increase in production costs.

Prolonged droughts, insufficient rainfall during wet years, and particularly temperature variability are perceived as having impacts on the seasonal cycle of arboriculture species.[100] Temperature peaks are becoming more frequent not only in the summer but also in the winter, leading to issues for the normal dormancy period for plants.[101] In the Gabès governorate, pomegranate production was cut by half in 2009, and some fruit trees, particularly apple, pear, and peach trees, skipped flowering periods altogether, also affecting yields.[102] The example of apple trees was also given in the Kasserine governorate as an illustration of the impact of temperature variability.[103] Due to higher-than-average temperatures in April 2011, flowers fell from the trees, resulting in a decrease in production.[104] This was also the case for the tomato crop: one generation in three was lost when flowers fell due to higher-than-average temperatures in the spring of 2011.[105] In Djerba, grapes and peaches from the region are usually the first to arrive on the national market, with an associated advantage of first entry.[106] With climate variability, however, flowering that normally occurs in the spring is occurring in autumn, leading to decreased production.[107] This variability has been exacerbated by sporadic, abnormal instances of frost and *sirocco* in Djerba.[108] The Gabès governorate also saw an increase in windy days during 2010, which damaged greenhouses, and from which small farmers are still recuperating.[109] Another issue raised generally has been the emergence of new diseases and insects, which have led to increased use of pesticides.[110]

Arid Areas of the Southern Governorates

In the arid southern governorates, rains usually come in the early spring and early fall.[111] There are also around 70–120 days of *vents de sable* or hot winds on average, of which 40 days are *sirocco*, especially during the late spring and summer.[112] During the rainy season, storms can bring significant amounts of rain within just a few hours, which can sometimes lead to flash flooding (120 millimeters in two days in 2007 for the Tozeur governorate, a reportedly one in a hundred-year occurrence).[113]

For date production in oases, supplemental irrigation is provided during the peak temperature summer months of July and August. Following the coming of the autumn rains in September or October, dates usually ripen (become *cuites* or "cooked") in November, when harvest takes place. In winter, the soil is worked, and in spring, artificial pollination takes place.[114]

As in the semi-arid areas of the central and southern governorates, there is a general sense that seasons have been altered and that the cycle of wet and dry years is now modified, with more consecutive dry periods punctuated by shorter and insufficient wet periods in between. With the general decrease and variability in rainfall, the cycle of date production is considerably altered, as the timing of rains in the cycle is crucial. Rains have not been coming regularly during the autumn period for the last 15 years, and dates are smaller and drier as a result. Additionally, rains come increasingly in the late summer and late spring. Coupled with the high summer temperatures, this causes the dates to "burn," further reducing yields. Heat peaks have prompted a prolonged period of irrigation through pumping from May to September (instead of July to August), with peaks in July and August. Wind patterns have also been changing, so that hot winds and *sirocco* come during an extended eight months of the year, leading to sand encroachment and direct damage to the dates. Another observation has been that of unusual temperature peaks between September and October (over 45°C)[115] combined with *sirocco*. This causes dates to dry prematurely in addition to damaging the fruit.

The lack of rain and increased temperatures has resulted in a general decrease in humidity within oases, threatening the traditional three-tier system. Farmers can no longer plant vegetable and fodder crops during the autumn with the assurance that rains will come, and the increased need for irrigation and associated costs means that these crops are the first to go from the three-tier system, followed by fruit trees.[116]

Consecutive droughts and rangeland degradation have also had a powerful impact on agro-pastoral livelihoods, because the normal transhumance cycle has been drastically altered. In the Kébili governorate, for example, agro-pastoralists used to rely on natural rangelands at the edge of the Sahara: their animals would stay near the oases to feed on vegetation wastes during the summer months[117] and once the autumn rains arrived, graze in the rangelands the rest of the year.[118] The present situation is quite the opposite: animals remain near the oases and rely on supplemental feed most of the year, due to the variations in rains and the quality of the rangeland, which is completely degraded from March to October, with peaks of degradation in August and September.[119]

Specific examples were also given for the impacts of seasonal variability on animal production in the Gabès governorate. With the prolonging of the summer period, fertility has decreased in cows, with lower fertility rates and increased embryonic deaths, especially over the last three to five years. Births in June and July see a 20 percent mortality death rate, even within the oases microclimates. Additionally, the lactation period usually lasts five to six months and is now reduced to three to four months, with direct impacts on revenues. The same trend is observed for goats, sheep, and to a certain extent camels.[120] Small animal raising, particularly rabbits and chickens, is also affected, with an increased death rate of young animals during the hot months. Farmers increasingly abandon this activity from June to August. It was also observed that apiculture production is declining, particularly in the Matmata area.[121]

Household Vulnerability to Climate Change

Three main categories of households emerged from the analysis of the field data. They are presented according to their vulnerability to climate change impacts.

- **Least vulnerable:** Families in this category had the following characteristics: (1) diversified sources of fixed income in addition to agricultural revenue (emigrated family member or one with a permanent job); (2) diversified agricultural products; (3) access to both water harvesting and irrigation; (4) land ownership; and (5) members of an older generation.
- **Moderately vulnerable:** This category covers families with only one of the "least vulnerable" characteristics, as these determine the length of time or number of droughts a family is able to cope with.
- **Most vulnerable:** Families in this category had a one or a combination of the following characteristics: (1) no fixed source of income; (2) reliance on only one agricultural product; (3) reliance on water harvesting as sole water resource; (4) no land ownership; and (5) unemployed young people. This category generally included small farmers.

Livelihood Strategies and Assets for Adapting to Climate Change

In general terms, the analysis showed that vulnerability is intrinsically related to the level of diversity of income sources available to the family, both within (in terms of diversity of products raised and sold) and outside agriculture. Vulnerability is also related to access to irrigated parcels and land ownership, which allows for a diversification of the water resource and production systems. The types and levels of vulnerability are, according to the IPCC definitions, related to adaptive capacity. In this section, we aim to capture the socially differentiated livelihood strategies, needs, and availability of assets that the rural population draws upon to respond to changing conditions. These conditions include natural resources affected by climate change impacts such as temperature variability, extreme events, long-term drought, and new precipitation patterns.

While the qualitative information and quantitative data below were systematically gathered, the aim was never to produce statistically significant results. This would have required a larger field operation than the 10-day field mission undertaken for this analysis. However, the qualitative information gathered is fully consistent with the quantitative data. The quantitative data have been ordered using the following "spiderweb diagrams" based on a series of concrete and ranked closed questions, targeted to reflect relative access to livelihood assets over time. The value 3 reflects the most favorable option to the interviewee and the value 1 the least favorable. For instance, 3 equals *high access to environmental resources such as water*, 2 equals *medium access*, and 1 equals *low access.* The colors of the lines connecting these perceptions of access indicate the present, past and near future. These diagrams should be used with the caveat that only 51 interviews were undertaken.

Deterioration of Environmental Resources

Water scarcity and overexploitation are central challenges for agricultural livelihood and agroecosystem sustainability and adaptability in the semi-arid and arid central and southern regions.[122] Agriculture uses most of the regions' water (82 percent), followed by households (13 percent), industry (4 percent), and tourism (less than 1 percent).[123] Even in a water demand scenario that does not take climate change impacts into account, conventional water resources will satisfy only an estimated 91 percent of demand by 2030.[124] The annual total volume of exploitable water resources in Tunisia is 4,800 cubic millimetres per year [125] or a quota of around 450 cubic meters per capita per year, which is set to fall to 360 cubic meters by 2030.[126] Increasing demand, coupled with climate change impacts, is projected to reduce already limited water reserves by 28 percent by 2030.[127] Coastal water tables and nonrenewable aquifers will be the most seriously affected water resources.[128] It is clear that water demand by 2030 and 2050 will be difficult to meet without the use of unconventional water resources or additional water conservation strategies.[129]

The central and southern regions once contained numerous natural water springs that fed oases, rangelands, and other agricultural lands. These springs have dried up over the last few decades due to decreasing rainfall and overexploitation, which have reduced the ability of aquifers to recharge.[130] Overexploitation of water resources stems in part from the extension of irrigation periods, particularly during summer months, that are needed to make up for irregularities in rainfall and temperatures.[131] The problem of groundwater depletion is especially salient for the southern parts of the central region and the arid southern region, where a higher proportion of nonrenewable fossil aquifers serve as the main water source. The depletion of these aquifers has created a sustainability issue for the entire economy of these regions.[132] In the Kébili governorate, the two main fossil aquifer systems are currently exploited at 205 percent and 228 percent of their respective capacities.[133] In the Tozeur governorate, decreasing water quantity and quality are being witnessed at well pumps, with outputs currently at 40 liters per second compared to 70 liters per second in the past.[134] Wells have to be drilled ever deeper (from an average of 0.7–1.5 meters at present), with associated increases in costs.[135] The increase in the number of wells over the last few years has led to competition between all sectors,[136] and a governorate-wide ban on new borehole wells.[137]

Water quality is also impacted by declining water resources through increased salinization.[138] In general, 88 percent of aquifers in the central and southern regions already have salinity levels above 3 grams per liter and in governorates such as Médenine, salinity ranges exceeds 5 grams per liter on average for most water resources (68 percent). The results are an associated salinization of the soil and quality problems for potable drinking water.[139]

Issues related to water resources, particularly access to water, often emerge as the defining environmental indicator of rural populations' capacity to cope with and adapt to climate change impacts. The semi-arid and arid regions of Tunisia are no exception. During group and individual interviews, the two issues that were stressed the most were water scarcity and the increased salinity levels of water at borehole wells. Many respondents in both semi-arid and arid areas tied migration issues to availability and quality of water. Furthermore, although respondents indicated that the distribution of water resources (borehole wells, irrigation, and so on) had in the past depended on access to political influence, many expressed the hope that the new administration will do a better in distributing and providing access to water resources in the future.

The field research indicates that while mean access to water was considered high in the past, it has now decreased to a medium level (figure 4.1). The picture, however, becomes more nuanced with disaggregated mean values. The rural population with access to oases management (figure 4.4) expressed negative sentiments with regard to access to water (good in the past, bad access now, expectations of medium access in the future). These respondents explained that there is now less humidity in the oases, partly due to increased pressure on water resources from enlarged oases and partly due to measures taken to use water more efficiently. For instance, canals are now in pipes, which do not release humidity into oases. These respondents also claimed that potential adaptive measures such as dams or water transfer have not been taken in time to deal adequately with climate change impacts. Ideas about potential solutions differed: some suggested deeper boreholes wells, while others were more inclined toward

Figure 4.1 Central, Southern, and Central Coastal Regions of Tunisia: Perceived Climate Change Management Capacity and Livelihood Assets

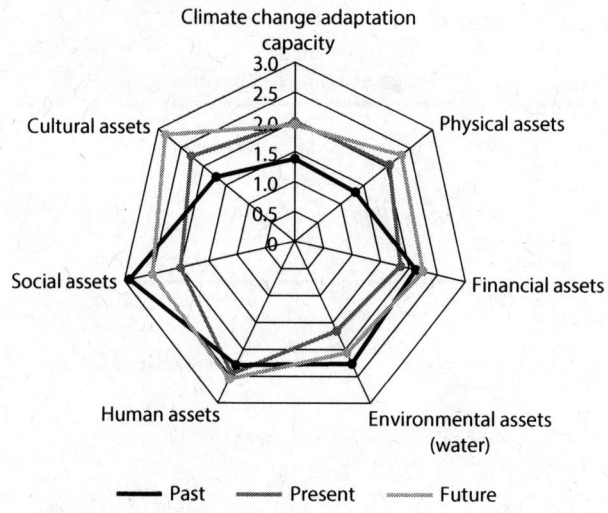

Source: Based on 51 interviews in 20 communities and sites, in the central and southern regions of Tunisia, November 2011. The figures are based on a series of concrete and ranked closed questions.

water transfer. Oases farmers are particularly concerned because a lack of water in the oasis results in increased costs and decreased production. As mentioned above, there is also a tendency for farmers to decrease crop diversity when faced with water scarcity, abandoning the traditional three-tiered oasis production system in favor of concentrating resources on date palms.

Many respondents without access to oases livelihoods explained that the lack of water is a condition that they have always lived with and adapted to. Even so, respondents in the arid southern region also maintained that access to water is worse now than in the past (figure 4.2). They perceived that there has been less rainfall in the last decade, apart from last year (2010). This variability, combined with greater pressure on water resources from increased agricultural

Figure 4.2 Disaggregated by Regions: Perceived Climate Change Management Capacity and Livelihood Assets

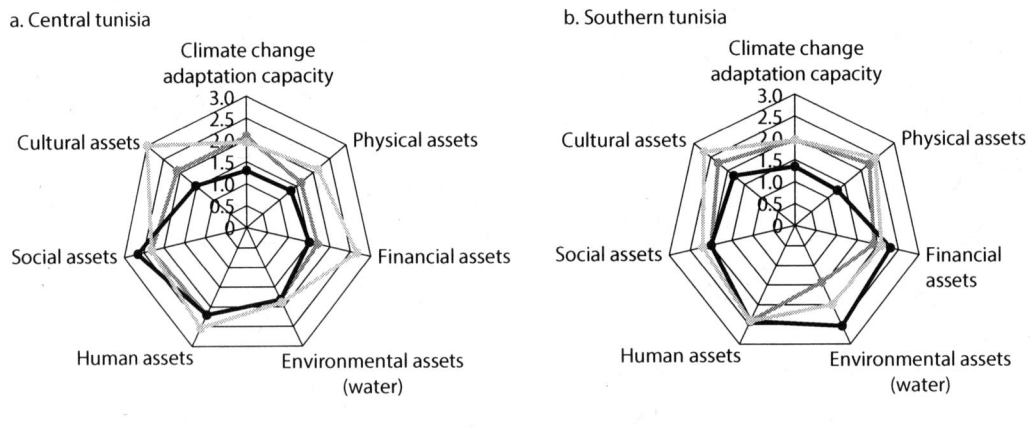

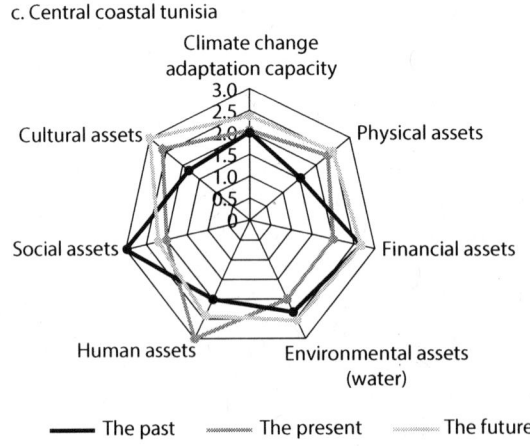

Source: World Bank data.
Note: Based on 51 interviews in 20 communities and sites in the central and southern regions, November 2011. The figures are based on a series of concrete and ranked closed questions.

activity, as well as private and public consumption, has meant that wells have had to be dug deeper and often become brackish.

In all regions and agroecological areas visited, respondents agreed that access to physical assets (infrastructure and transport of people, goods, animals, fodder, and water) has improved over past decades (ranked from bad to medium, figure 4.3). In the past, rural populations relied mainly on animal transport, with conditions improving as a result of the use of vehicles and rural roads. However, respondents did rank rural roads as declining from medium to bad, as they felt that road maintenance has been affected by flooding, which has in turn impacted prices of food, fodder, water, and human transport due to more difficult access. In several of the visited communities, there was no adjacent public transport other than the school bus. The issue of poor infrastructure is seen as clearly linked with unemployment. Better roads would increase availability of jobs locally, expand mobility, and keep people from permanently migrating to bigger cities.

Decreasing Access to Financial Capital

Access to loans, credits, and various sources of income is limited and declining in central and southern Tunisia. Farmers who once had productive three-tiered oases with a diversity of crops now harvest only from date palms. Yields have diminished and alternative sources of income are scarce. Farmers with olive tree plantations produce less due to dry years and abstain from complementing their income with vegetable and livestock production. For agro-pastoralists, increased dependence on supplemental feed has increased the costs of production and decreased profitability. Several respondents further emphasized that the young have no, or very limited, access to funds. This situation is clearly

Figure 4.3 AgroEcological Zones: Perceived Climate Change Management Capacity and Livelihood Assets

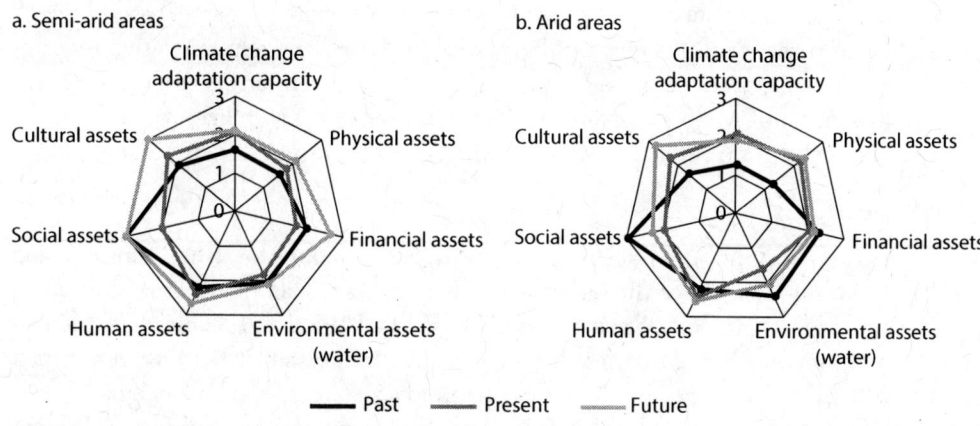

Source: World Bank data.
Note: Based on 51 interviews in 20 communities and sites in the central and southern regions, November 2011. The figures are based on a series of concrete and ranked closed questions.

**Figure 4.4 Specific Production Conditions: Perceived Climate Change Management
Capacity and Livelihood Assets**

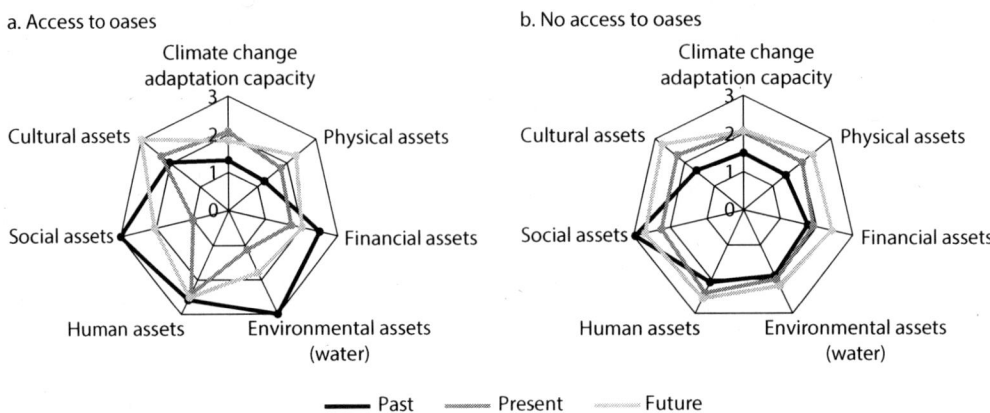

Source: World Bank data.

Note: Based on 51 interviews in 20 communities and sites in the central and southern regions, November 2011. The figures
are based on a series of concrete and ranked closed questions.

reflected in figure 4.4, which shows a decline in access to financial assets (high
to medium) coinciding with a drastic decline in both social assets (trust) as well
as environmental assets (water). In fact, all respondents answered that they had
high access to financial capital in the past and low access at present. Although
respondents in general linked the state of future production to availability of
water and an efficient postrevolutionary government, people of the central
coastal region (Djerba) mentioned that they will aim get involved in or increase
existing ties with the tourism industry, and to a limited extent fisheries.

In central and southern Tunisia, many respondents mentioned that although
small farmers generally have difficulty getting loans, those with access to irrigation had more success, as they could use their irrigated parcels as collateral.
Interest rates are perceived as being very high and the administrative requirements for getting a formal loan as very stringent. Even farmers who can obtain
loans try not to do so, because of the uncertainty in the reliability of access to
water and other resources, which might make repayment difficult or impossible.

Challenges to Social Ties and Networks

Formal and informal institutions, networks, relations of trust, kinship, and
friendship constitute the "glue" of rural communities and have crucial implications for livelihoods. Most respondents ranked their current level of trust as
medium, indicating that it is worse now than in the past or the foreseen future.
The most pronounced difference was found among respondents who had access
to oases, as reflected above in figure 4.4. In this analysis, we define social assets,
or social capital, as rules, norms, obligations, reciprocity, and trust embedded in
social relations, social structures, and societies' institutional arrangements. At

present, social assets among those with access to oases are at the lowest possible level (value equal to 1), while the same group of respondents agreed that there was a high level of trust and good social relations one generation ago. Respondents explained this lack of social capital by way of shame and uncertainty. They indicated that they would not ask friends for loans because they felt ashamed, because they would not be able to pay back the loan, or because few people have anything to lend. Other respondents indicated that they still had access to such loans from family and friends through their networks.

The present is referred to with much uncertainty, and it becomes clear that social, financial, and environmental livelihood assets are closely connected, and relevant to how people respond to changing conditions. Interestingly, respondents in general indicated some hope for the future with the new postrevolutionary environment. Box 4.9 provides anecdotes from individual interviews, illustrating the issue of changes to social ties and networks linked to climate variability and change.

Migration of Youth and Changing Gender Roles

The special knowledge and experience that people acquire from living under certain conditions, and by relating to each other over time, are among the

---

Box 4.9

Challenges to Social Ties and Networks According to Respondents in the Central and Southern Regions

On the issue of changes in the level of trust, an oasis farmer in Hazoua Old Oasis in the Tozeur governorate said:

> Yes, the drought has affected trust in the community. People look for stability and this variability is worrisome. For the family, members migrate and this affects the whole community. The growing demand for water will continue. Before, with the nomadic way of life, it was a completely different situation. Clear goals can bring a community together, but we need attractive options.

Two respondents from the Tataouine governorate in the southern region observed:

> With rain, it's a party. When drought comes, there are negative impacts and problems arise among communities. People are stressed, worried, ways of life change…. Excessive drought affects young people who migrate to other regions, especially the north. There is a decrease in the quality of relationships among people.

Likewise, a respondent from Djerba commented: *The changing climate does affect relationships. Those with means can continue to work. Those who don't [have means] can't. This creates inequalities in the capacity to save during drought. Trust was much better before. Before, you could just do transactions based on verbal agreement. The word was more valuable than money.*

Note: Based on individual interviews in the central and southern regions in November 2011.

cultural dimensions of livelihood strategies. Ideas about identity and differentiation from others often come out strongly in interviews about these issues. In the semi-arid and arid central and southern regions, the most significant cultural dimension referred to was that of age, or generation. This is perhaps unsurprising, given that the revolution had taken place only months before and highlighted the current difficult situation for Tunisia's youth. Most respondents agreed that young people cannot and will not live as their parents have, with many young people now fleeing agriculture and seeking a better quality of life through temporary jobs in the city. Nonetheless, some remain, drawing on cultural practices, networks, and knowledge that may enhance their capacity to adapt. As a young man from the Gafsa governorate said:

> I have the knowledge to safeguard my plantations and livestock. I count on my savings from rainy years to get through drought. I have knowledge that even experts don't have on water flow, types of species—traditional knowledge— and can pass this on as it was passed down to me. City people don't have this, but also don't have the same worries I have about my property and crops and livestock.

The young man continued on to say that his father wanted things to stay the way they are. He himself wanted things to change, and to look for new production techniques. Young people, in his view, are more active and curious to try new things.

Other respondents referred to gender relations and increasing levels of education to explain changes in the culturally based production systems of their community. As one respondent from the Kébili governorate noted:

> Before, women practiced 80 percent of agriculture. Now, they refuse to do agriculture and manual labor. They are educated now. Knowledge and know-how have increased. But people won't keep doing agriculture, if conditions are not good.

Figure 4.5 Perceived Climate Change Management Capacity and Livelihood Assets

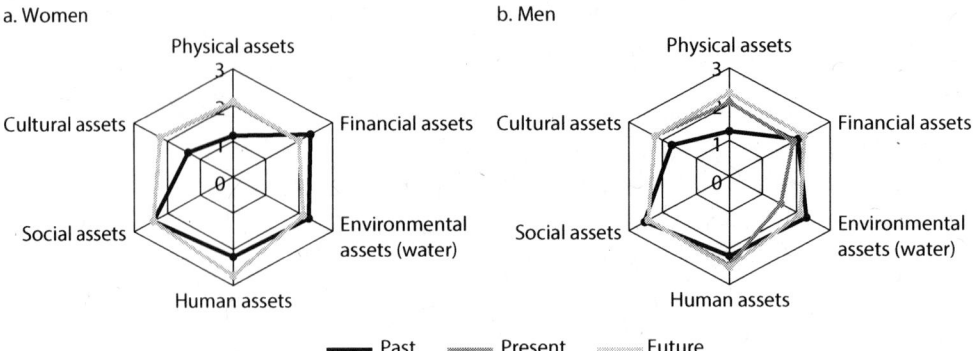

Note: Based on 51 interviews in 20 communities and sites in the central and southern regions, November 2011. The figures are based on a series of concrete and ranked closed questions.

Human Assets

From Figure 4.5 we learn that men and women express quite similar access to livelihood assets over time with a couple of differences. Men are more positive regarding physical assets for the future than the women but currently express less access to water. Women have higher expectations to future access to human assets such as health and education. Human assets include health, education, and nutrition. Education levels are relatively high in the rural households of central and southern Tunisia. Most respondents have completed primary and secondary education, and most families have members, often female, with higher levels of education. Such knowledge increases the ability to access climate-related information and relevant government information, and opens the door to other sources of income in urban and semi-urban sectors. This sustains the widespread hopeful expectations of many respondents of the new government. Access to private and public health care, and to a varied diet, depend largely on financial means, and is generally ranked between good and very good by respondents. While the general level is good and does not seem to be changing, female respondents coincided in saying that their access to health and particularly education assets are improving (see figure 4.5).

Operational Recommendations for Adaptive Responses and Institutional Measures

This chapter has examined the key socioeconomic implications of climate change and variability for the populations of selected communities in seven governorates of central and southern Tunisia. Throughout the communities interviewed, there was a general sense that the new sociopolitical environment in the country would lead to an overall improvement in their way of life, in part resulting from a better, more equitable dialogue between themselves and national and local government. Communities and local government officials emphasized very strong links between regional development and territorial planning, particularly regarding access to basic infrastructure and capacity for adaptation. A sound working relationship between communities and key local government actors, including agricultural extension service agents, is paramount to building adaptive responses that address specific community vulnerabilities to climate change impacts.

Five principal conclusions can be drawn from the analysis:

1. *Impacts Are Multifaceted and Differentiated:* Food production systems and the agroecological conditions sustaining local livelihood strategies are severely stressed. Historical overexploitation of water and soil, coupled with climate-related stress on living and production conditions, has negative impacts on rural income generation and employment; food security, both at the household and national levels; and natural resources.

2. *Local Actors, Governments and Sector Institutions Need Support to Implement Identified Actions:* Given the new political environment in the country and the heightened expectations of local communities, there is a need for a systematic,

Table 4.1 Recommendations for Improving Resilience and Adaptability of Rural Communities in the Central and Southern Regions

Recommendation	Measures
1. Identify and implement adaptation measures that synergize traditional local knowledge with new techniques. A participatory approach should be used involving members of the older generation, agricultural extension technicians, and agriculturalists (particularly young people). Material incentives (inputs, etc.) to promote successful techniques should be provided.	Identify (through a questionnaire or other means) spontaneous adaptive measures and techniques already undertaken by agriculturalists when faced with climate change impacts, to be shared across communities. Promote more water-efficient and drought-resistant fodder species, such as acacia and cactus,[a] and other native species to guarantee fodder reserves during dry years.[b] Carry out adaptive animal health programs, including veterinary services, to decrease pressures on vulnerable natural rangelands. Restore traditional three-tier oases systems, as the microclimate provides a buffer against temperature and rainfall variability, and pilot new techniques to restore humidity in the oasis (spacing of date palms, etc.) Promote more water-efficient and drought-resistant arboriculture species (olive, pistachio, and almond trees) in areas that can resist prolonged droughts. Pursue interventions to stabilize sand dunes to protect oases, crops, and infrastructure. Pilot techniques to improve the quality of soils and their ability to retain water, using experiences in organic agriculture as a guide and avoiding extensive fertilizer use. Expand soil and water conservation measures based on successful traditional designs, such as *tabbias* and *jessours*, and innovative techniques such as the *poches en pierre*.[c]
2. Improve water management through a cooperative approach to transboundary aquifer systems, increased efficiency, and the use of alternative water resources. For transboundary aquifer systems, completed work on a transnational approach to the management of the southern region's main aquifer systems[d] should be built upon. A culture of efficient water use and management should be disseminated among main water users. The full range of runoff and alternative water resources should be explored.	Sensitize large water users[e] (the agriculture and industrial sectors) and individual water users to the need for efficient use of water, building on existing incentive programs (60 percent subsidies on water-efficiency projects) and the National Society for the Use and Distribution of Water (SONEDE)[f] water saving strategy for domestic users. Preparing a best practice guide may be useful. Optimize the efficiency of new and existing wells by, for instance, placing new wells strategically and mapping out the location of existing wells to be rehabilitated or closed. Expand soil and water conservation measures to collect runoff water through hill lakes and water-spreading techniques from *oueds*, using modified designs resistant to flooding, hail storms, and other extreme events related to climate change.[g] Pilot soil and water conservation works to specifically increase soil resistance to water and wind erosion, improve the infiltration capacity of soils (small-hydraulics systems),[h] and allow for the recharge of groundwater aquifers. Examples include systems to retain seasonal floodwaters on large *oueds*[i] that use both traditional and modern techniques. Pilot the use of alternative water resources such as irrigation with drainage and brackish waters (up to 10g/L of salinity), desalination options, and treated wastewater of a quality acceptable to agriculturalists. Employ tertiary treatments options where needed.

(table continues on next page)

Table 4.1 Recommendations for Improving Resilience and Adaptability of Rural Communities in the Central and Southern Regions *(continued)*

Recommendation	*Measures*
3. Support the diversification of sustainable revenue sources for rural populations to decrease their vulnerability. Communities should be supported in developing alternative income sources, both non-agricultural and agricultural, with sufficient capacity to commercialize products. Agricultural income needs should be balanced against water availability and resource sustainability.	Explore the full use of alternative fodder subproducts (acacia, olive wastes, and wood wastes).[j] Explore the development of seaweed culture (spirulina) along the coasts as appropriate (Djerba and elsewhere). Develop revenue opportunities that work in tandem with sustainable management of protected areas (ecotourism). Enhance credit schemes for young people who are willing to work in agriculture and support their communities using innovative techniques.
4. Design and implement tailored decision support and capacity-building programs for local communities and agricultural extension services to carry out identified adaptive measures. Initiatives should be tailored to the technical knowledge of the targeted agriculturalists and rural population. Technical support should be reinforced for agricultural extension services, the first "links" between local government and the local population. Actions taken on the ground must then be monitored and evaluated through recurrent contact with the local population and assessment of lessons learned.	Raise awareness about the vulnerability of monocultures and promote integrated and diversified agricultural production systems. Enhance the capacity of agro-pastoralists to build up fodder reserves during wet years and adapt their practices to new rainfall patterns.[k] Promote mechanisms for community engagement in traditionally "unpopular" adaptive methods, such as the cutting of olive trees and reseeding rangeland reserves with native species. Raise awareness and provide training on identified climate change themes for CRDAs and NGOs, to fully sensitize communities.
5. Institutionalize lessons learned. Importance should be given to initiatives with impacts for long-term planning impacts by local government actors and balanced with the immediate need of local communities. Tools for decision support at the regional level should be developed, to integrate uncertainty into planning. A broader set of actors should also be involved, particularly NGOs linked to environmental and sustainable development, research institutions, and universities.	Develop practical intervention options based on different wet and dry year scenarios to be implemented by CRDAs. These options should take into account climate variability and different agroecological zones within a given governorate to increase readiness to subsequent droughts.[l] Use downscaled climate modeling to measure the dynamic impacts of climate change on specific production systems for planning at the regional level. Establish national and regional observatories as well as a national observatory with regional capacity to ensure a scientific approach to monitoring and measurements of climate change. Fulfill research needs on the following themes: – Linkages between climate change and the appearance of invasive species, new diseases, and new parasites. – Linkages between climate change and fisheries health. Review the organization and representation of Agricultural Development Groups (GDAs) in the new sociopolitical environment, so that local populations are comfortable with GDAs as their representatives vis-à-vis the local administration and agricultural extension services.

a. 83,000 hectares of cactus currently planted in the Kasserine governorate.

b. Central region. *Tabia* and *jessours* are traditional types of soil and water conservation. A *tabia* is a type of small earth bund and a *jessour* represents the resulting water accumulation behind the *tabia.*

c. Southern region. *Poches en pierre* or "stone pockets" is a type of water harvesting and underground irrigation technique involving several rows of stones placed at the end of a small-scale "canal" between rows of crops/trees.

d. The *Continentale Intercalaire* aquifer shared by Algeria, Tunisia, and Libya.

e. For example, the Gafsa governorate has three main water users: the mining industry, Gafsa city, and the oases.

(table continues on next page)

Table 4.1 Recommendations for Improving Resilience and Adaptability of Rural Communities in the Central and Southern Regions *(continued)*

Recommendation	Measures

f. Société Nationale d'Exploitation et de Distribution des Eaux.

g. Central region. Especially important in mountainous areas where production losses can be up to 40 percent due to erosion from rains.

h. Especially important for the southern region considering that the water needs of an oases parcel is based on the ***tour d'eau*** or water turn.

i. There are three barrages currently in the Gafsa governorate to capture large floodwaters and protect the Gafsa plain. The semi-underground barrage Sidi Boubaker in the north is specifically aimed to recharge the aquifer. Starting in 2012, a new barrage on the Oued Ekbir will contribute to groundwater recharge and protecting Gafsa city against flooding.

j. Central region.

k. Central region.

l. In the Gabès governorate, olive trees are being increasingly introduced away from the Mediterranean climate zones, resulting in a vulnerable placement, and fruit trees are planted without knowledge of the suitability of the soil.

general focus on "low-hanging fruit" adaptive options (see examples of these measures in table 4.1 below). The new government, however, faces what a high-level official calls a "triple uncertainty." The administration must currently address a fluid post-revolution political and institutional environment; it must cope with climate variability and change, and the depletion of critical natural resources; and it must deal with the financial crisis in Europe, which Tunisia depends on for exports and tourism.

3. *Uncertainty Should Be Systematically Integrated into Planning:* Multiple and often related uncertainties will remain key concerns in future planning and implementation of sustainable agricultural production systems. The effects of shocks, extreme events, and the variability and unpredictability of weather patterns should be integrated into strategies to conserve and restore depleted agroecological resources, and to adapt to long-term climate change trends of drought and precipitation variability.

4. *Local Livelihood and Production Options Should Be Diversified:* Providing people who live off the land with viable alternative livelihoods is especially important in order to preserve social structures, by reducing migration and absenteeism. Further, their experience and knowledge of the land in itself represents a tool for adaptation. The expansion and continuity of traditional (for example, dates and olives) and alternative (for example, cactus, native fodder species) agricultural products will need to be balanced against the availability of water and the capacity of these species to resist drought. Regarding potential adaptive responses, two points were stressed during interviews in particular: (1) the importance of an integrated management approach to agro-ecosystems and associated natural resources, particularly water and soils; and (2) the need to diversify production systems, providing communities with alternative income sources. Boxes 4.10 and 4.11 present examples of successful integrated agro-ecosystem management and diversification of agricultural revenue sources for rangelands and oases.

5. *Investment in Research, Dissemination of Information, Awareness Raising, Skills Training and Capacity Building are all Needed*: In general, water scarcity and extreme weather events caused and exacerbated by climate change

Box 4.10

Rangeland Rehabilitation with Native Species

In the Douz area of the Kébili governorate, the Minisitry of Agriculture has undertaken a rangeland rehabilitation program to increase rangeland agro-ecosystem and associated pastoral livelihood resilience to degradation pressures. With financing from the International Fund for Agricultural Development (IFAD), a 15,804 hectare semi-agro-pastoral zone has been set aside for revegetation by native desert plants. The process begins with sowing seeds developed in nurseries in the Center for the Production of Saharian Seeds/*Centre de Production de Semences Sahariennes* in Kébili, a first for Tunisia, created in the context of the Program for Agro-pastoral Development and Promotion of Local Initiatives in Southeast Tunisia/*Programme de Dévelopment Agropastoral et Promotion des Initiatives Locales dans le Sud-Est Tunisien* (PRODESUD). Rehabilitation areas are set aside for revegetation based on the native plants' natural cycle.

The project is based on a participatory process with advice from community groups. It benefits GDAs, which previously did not have sufficient operational funds to service their constituents.

A major challenge to this initiative is water scarcity affecting plant viability and reproduction. Native desert plants forgo seed production during droughts, and so need some watering in February, in order to produce seeds in May. Excess water that washes away vegetation during short-lived but intense flooding during the rainy season represents the other side of the water resource challenge.

Rangeland Undergoing Rehabilitation with Native Species in the Kébili Governorate

Source: Viviane Clément.

(box continues on next page)

Box 4.10 Rangeland Rehabilitation with Native Species (continued)

To address these issues and enhance durability, the project is trying to make the most of the natural characteristics of both native plants and their environment. For instance, the native *tamaris* plant, adapted to saline water and soils, can be used to mitigate erosion and serve as natural barriers to sand encroachment. In terms of water use, the project promotes cultivation measures adapted for arid soils, such as covering seeded areas after watering with sand to prevent evaporation and preserve moisture, or taking into account the *oued* beds' inherent moisture. The project has also incorporated small water points for native birds, which act as natural seed dispersers. One of the additional benefits of the project has been the return of native biodiversity with revegetation efforts.

Source: Based on field visits to the southern region, February 2010.

Box 4.11

Integrated Oasis Management: Example of the Hazoua Oasis

In Hazoua in the Tozeur governorate, the Biodynamic Hazoua GDA (grouping 3 small farmer GDAs) is taking an integrated approach to oasis production and management, as well as that of the surrounding rangelands. The three-tier cultivation system is being revived here to conserve moisture and biodiversity in the oasis, while maintaining production and quality levels. Dates are destined for export to Germany and Scandinavia, and are certified organic and fair trade.[a]

To resolve water scarcity and soil health issues, new techniques are being tested, namely integrating legumes and seedlings into the parcels, as well as small hydraulics irrigation systems (localized submersion and drip irrigation) that are adapted to sandy soil type, and the use of PEHD pipes.[b] The initiative also exports the produced seedlings, also certified organic.

The initiative is also testing a number of protection measures against *sirocco* winds and pests, including wind breakers, bags around date clusters, vegetation enhancement, and integrated pest management.

The initiative also includes 200 hectares of improved set-aside rangelands. Healthy surrounding rangelands serve as a natural protective measure for the oasis, acting as a belt. Over 20 hectares of the rangeland, drainage waters (8–12 grams per literg salinity) from the oasis (using solar pumping) are being tested for irrigation. Results are promising with both perennial and annual native species being rehabilitated, as well as barley, thus allowing for the full use of the water resource potential. The set-aside area is protected by guards and a solar electric fence. The fodder produced is used not only as traditional fodder, but also as compost to enrich the soils in the oasis, as the species grown are rich in nitrogen. Vegetation wastes from the oasis is also itself used for fodder. The integrated management of the oasis and rangeland has also allowed for the regeneration of native fauna, including native hares and foxes.

(box continues on next page)

Box 4.11 Integrated Oasis Management: Example of the Hazoua Oasis *(continued)*

Integrated Oasis Management in Hazoua, Tozeur Governorate

Source: Jakob Kronik.

Source: Field visit to the Hazoua oasis in November 2011.
a. 18 parcels were certified in 1992, and other parcels have been certified since 2000, through Bio Suisse, European Union (EU), Demeter International, and the USDA. Processing certification took three years on average and certification maintenance costs average 20,000 TD per year.
b. Cost is around 10,000 TD per hectare.

require an actionable, region-specific agricultural adaptation strategy. Climate and weather projections, coupled with early warning systems, must include timeframes that are useful in predicting harvests for farmers. Rural populations must have access to timely, relevant, and reliable information about climate change phenomena, related impacts and targeted information to strengthen context-specific adaptation strategies, and the capacity to implement these strategies through tailored awareness raising and training programs.

The main recommendations from this chapter, to improve the resilience and adaptability of rural communities to climate change impacts in the central and southern regions of Tunisia, are organized in table 4.1. Examples of potential adaptive responses originate from local communities themselves and the CRDAs of the seven governorates visited for this analysis. Options are generally valid for both regions unless specified.

Notes

1. Ministère de l'Environnement et du Développement Durable/PNUD (2009, 20–21).

2. Ministère de l'Agriculture et des Ressources Hydrauliques/GTZ (2007, 5), Cahier 2.

3. Ministère de l'Environnement et du Développement Durable/PNUD (2009, 23).

4. Ministère de l'Agriculture et des Ressources Hydrauliques/GTZ (2007, Synthesis, 7).

5. Ibid., 7.

6. Ibid.

7. Ibid.

8. Ibid.

9. Ministère de l'Environnement et du Développement Durable/PNUD (2009, 23).

10. National Statistics Institute (2004) Census.

11. Ibid.

12. Ibid.

13. Ministère de l'Environnement et du Développement Durable (DGEQV)/FEM/PNUD (2008).

14. Bigio (2009). This report includes analysis on the cities of Alexandria, Casablanca, and Tunis.

15. Ongoing GIZ study on climate change impacts on the oak forests in the North region and World Bank Fourth Northwest Mountainous and Forested Areas Development Project (PNO4) and Second Natural Resources Management Project (PGRN2).

16. *Commissariat Régional au Développement Agricole*

17. *Groupement de Développement Agricole*

18. Ministère de l'Agriculture et des Ressources Hydrauliques/GTZ (2006, 187).

19. Ibid., 188.

20. Ibid.

21. Ibid.

22. United Nations Environment Program (2006).

23. Interviews with the Gafsa, Médenine and Tataouine CRDAs, November 2011.

24. Ministère de l'Agriculture et des Ressources Hydrauliques/GTZ (2007, 19), Cahier 3.

25. Ibid., 19.

26. Ibid., 20.

27. Interview with the Kasserine CRDA, November 2011.

28. Ibid.

29. Interview with the Gabès CRDA, November 2011.

30. Ibid.

31. Ibid., 21.

32. World Bank (2010).

33. From interview with the Médenine CRDA in November 2011.

34. Ministère de l'Agriculture et des Ressources Hydrauliques/GTZ (2007, 21), Cahier 3.

35. Ibid., 20.

36. Ibid.

37. From interviews with the Kasserine, Gafsa and Médenine CRDAs, November 2011.

38. Ibid.

39. Sold at 3TD and re-bought at 7-12TD in the Kasserine governorate. From interviews with the Kasserine CRDA, November 2011.

40. Sold at 20–30 TD per head. From interviews with the Kasserine and Kébili CRDAs, November 2011.

41. From interviews with the Kébili and Médenine CRDA, November 2011.

42. From interviews with the Kasserine, Gafsa and Médenine CRDAs, November 2011.

43 Ibid.

44. From interview with the Tataouine CRDA, November 2011.

45. 240,000 hectares.

46. Three million out of 5.38 million date palms.

47. Seventy-three percent of date palms are of the *Deglet Nour* variety in the governorate.

48. From interview with the Kébili CRDA, November 2011.

49. Notes from field visit to the Gabès oasis, February 2010.

50. Ibid.

51. Ibid.

52. UNESCO (2010).

53. Field visits to the South region, February 2010 and November 2011.

54. From interviews with the Gabès and Gafsa CRDAs, November 2011.

55. From interviews with the Tozeur CRDA, November 2011.

56. Ibid.

57. From individual interview in Maksem Echik community, Tozeur governorate, November 2011.

58. From interview with GDA in Hazoua, Tozeur governorate, November 2011.

59. Ibid.

60. Ibid.

61. Ibid.

62. From notes from field visits to the South region, February 2010.

63. Of which 1.3 million ha is olive tree monoculture and 300,000 ha of olive tree plantations are associated with other arboriculture trees. At the national level, 32 percent of olive trees are less than 20 years old and 26 percent are over 50 years old. (Ministère de l'Agriculture et des Ressources Hydrauliques/GTZ (2006, 182).

64. Ibid., 182.

65. Ibid., 183.

66. Ibid.

67. Ibid.

68. Ibid.

69. Field visit to the central and souther region, November 2011.

70. Field visit to the Gabès governorate, February 2010.

71. From interview with the Gafsa CRDA, November 2011.

72. From interview with the Tataouine CRDA, November 2011.

73. From interview with the Gabès CRDA, November 2011.

74. From interview with the Médenine CRAD, November 2011.

75. 75,000 tons per year production. Ibid.

76. Ibid.

77. From interview with the Gabès CRDA, November 2011.

78. From interview with the Kasserine CRDA, November 2011.

79. From interview with the Djerba CRDA representation, November 2011.

80. Ibid.

81. From interview with communities in Kasserine, November 2011.

82. Ibid.

83. From interview with communities in Tatatouine, November 2011.

84. Ministère de l'Environnement et du Développement Durable (DGEQV)/FEM/ PNUD (2008, 38).

85. Ibid., 38.

86. Ibid., 39.

87. From field visit to the Gulf of Gabès area, February 2010.

88. Ministère de l'Environnement et du Développement Durable (DGEQV)/FEM/ PNUD (2008, 39).

89. From interview with the Houmet Souk Port Authority, November 2011.

90. Ibid.

91. Ibid.

92. Ibid.

93. From interview with the Gabès CRDA, November 2011.

94. From notes from field visit to the Gulf of Gabès area, February 2010.

95. From interview with the Houmet Souk Port Authority and Gabès CRDA, November 2011.

96. From interview with the Houmet Souk Port Authority and Gabès CRDA, November 2011.

97. Field visit to the central and southern regions, November 2011.

98. For example 400 millimeters in 24 hours in one instance in the Gafsa governorate. From interview with the Gafsa CRDA, November 2011.

99. Field visit to the central and southern regions, November 2011.

100. From field visit to the Center and South regions, November 2011.

101. From interviews with the Gabès, Kasserine and Djerba (Médenine) CRDA, November 2011.

102. From interview with the Gabès CRDA, November 2011.

103. From interview with the Kasserine CRDA, November 2011.

104. Ibid.

105. Ibid.

106. From interview with the Djerba (Médenine) CRDA representation, November 2011.

107. Ibid.

108. Ibid.

109. Paragraph from field visit to the central and southern regions, November 2011.

110. Interviews with the Gabès, Kasserine and Djerba (Médenine) CRDA, November 2011.

111. Ibid.

112. Ibid.

113. Interview with the Tozeur CRDA, November 2011. There was also flooding in 2009 in Tamerza.

114. Field visit to the central and southern regions, November 2011.

115. Interview with the Kébili CRDA, November 2011.

116. Ibid.

117. Three months June-August.

118. From interview with the Kébili CRDA, November 2011.

119. Ibid.

120. There are 1,000 camel heads in the Gabès governorate.

121. Paragraph from interview with the Gabès CRDA, November 2011.

122. Tunisia's water reserves are made up of 51 percent surface water and 49 percent groundwater, with water resources exploited as deep aquifers (46.3 percent), shallow aquifers (33.2 percent), and dams (20.5 percent). Ministère de l'Agriculture et des Ressources Hydrauliques/GTZ (2007), Cahier 2.

123. Ministère de l'Agriculture et des Ressources Hydrauliques/GTZ (2006, 72), Final Report.

124. Ministère de l'Agriculture et des Ressources Hydrauliques/GTZ (2007, 16), Cahier 3.

125. Gaaloul (2011).

126. Ministère de l'Agriculture et des Ressources Hydrauliques/GTZ (2007, 16), Cahier 3.

127. Ibid.

128. Ibid., 11, Cahier 3.

129. Ibid., 16, Cahier 3.

130. From interview with the Kasserine and Médenine CRDAs, November 2011.

131. Ibid.

132. The *Complexe Terminal* and *Continentale Intercalaire* aquifers. From interview with the Gabès, Gafsa, Kébili and Tozeur CRDAs, November 2011.

133. *Complexe Terminal* aquifer exploited at 9,237 liters per second compare to a 4,500 liters per second capacity. *Continentale Intercalaire* capacity at 1,000 liters per second. From interview with the Kébili CRDA in November 2011.

134. From interview with the Tozeur CRDA, November 2011. Geothermal aquifers fulfill the needs of 90 percent of all economic sectors, 6 percent of industry needs, and 4 percent of potable water needs.

135. Ibid.

136. From interview with the Médenine CRDA, November 2011.

137. From interview with the Tozeur CRDA, November 2011.

138. From interview with the Médenine CRDA, November 2011.

139. From interview with the Djerba CRDA representation, November 2011.

References

Bigio, A. 2009. "Adaptating to Climate Change and Preparing for Natural Disasters in the Coastal Cities of North Africa." World Bank Report, World Bank, Washington, DC. http://siteresources.worldbank.org/INTURBANDEVELOPMENT/Resour ces/336387-1256566800920/6505269-1268260567624/Bigio.pdf.

Gaaloul, N. 2011. "Water Resources and Management in Tunisia." *International Journal of Water* 6 (1/2): 92–116.

Ministère de l'Agriculture et des Ressources Hydrauliques/GTZ. 2006. Elaboration d'une stratégie nationale d'adaptation de l'agriculture tunisienne et des écosystèmes aux changements climatiques. Rapport Final.

———. 2007. Stratégie nationale d'adaptation de l'agriculture tunisienne et des écosystèmes aux changements climatiques—Cahiers 1–7. Janvier 2007.

Ministère de l'Environnement et du Développement Durable (DGEQV)/FEM/PNUD. 2008. Etude de la Vulnérabilité Environnementale et Socio-économique du Littoral Tunisien Face à une Elévation Accélérée des Niveaux de la Mer Dues aux Changements Climatiques et Identification d'une Stratégie d'Adaptation. Mars 2008.

Ministère de l'Environnement et du Développement Durable/PNUD. 2009. Etude l'élaboration de la seconde communication nationale de la Tunisie au titre de la Convention Cadre des Nations Unies sur les Changements Climatiques—Phase III: Vulnérabilité de la Tunisie face aux changements climatiques. Juin 2009.

National Statistics Institute. 2004. "Population Census." http://www.ins.nat.tn/indexfr. php.

UNEP (United Nations Environment Program). 2006. "Global Deserts Outlook." United Nations Environment Programme, Nairobi, Kenya. http://www.unep.org/ geo/gdoutlook/.

United Nations Educational, Scientific, and Cultural Organization (UNESCO). 2010. "Oasis de Gabès." Paris, France. http://whc.unesco.org/en/tentativelists/5386/.

World Bank. 2010. "Tunisia—Second Natural Resources Management Project." Project Appraisal Document, World Bank, Washington, DC. http://www.worldbank.org/proj-ects/P086660/tunisia-second-natural-resources-management-project?lang=en.

CHAPTER 5

Recommended Climate Policy Responses and Actions in Tunisia

Photograph by Dorte Verner

This chapter introduces a framework for climate change decision making and provides concrete policy, plan, and program recommendations to address climate change in Tunisia. The proposed actions align with the World Bank's 2012 Interim Strategy Note (ISN) for Tunisia, which guides World Bank investment in Tunisia over the next two years. The ISN focuses on three main areas of intervention: (1) renewed sustainable growth and job creation, (2) the promotion of social and economic inclusion by improving access to basic services for underserved communities and improving the efficiency of social safety net programs, and (3) strengthening governance through improved access to public information as the basis for increased social accountability and transparency. The specific recommendation in this chapter are related to five main themes: (1) improving the quality and accessibility of public information related to climate change; (2) improving human, technical, and other resources and services to support climate change action; (3) providing social protection and other measures to ensure that the poor and the most vulnerable are climate resilient; (4) developing a supporting policy and institutional framework for climate change; and (5) building capacity to generate and manage revenue and to analyze financial needs and opportunities associated with adaptation.

This chapter builds on the analysis carried out in the first four chapters of the report as well as on the current sociopolitical situation in Tunisia elaborated in chapter 1. From the preceding chapters it is clear that climate change will result in increased temperatures, reductions in rainfall, and increases in extreme events such as droughts, storms, and floods in Tunisia (see chapter 2). The already-water-scarce country will become increasingly water scarce, threatening capacities to irrigate crops, support industry, or provide drinking water to all. In addition it is likely that significant climate impacts will occur as a result of changing world food prices, especially since Tunisia is a net importer of many food commodities. As elaborated in chapter 3, world market prices for food are projected to increase under climate change and local climate change impacts will manifest themselves through long-term yield changes. Yields for wheat, barley, and irrigated potatoes are expected to fall. Global (higher global food prices) and local effects (lower yields) together are projected to reduce economic output in Tunisia by between US$2 and US$2.7 billion over 30 years. Some farmers may benefit from the higher world food prices, but the overall effects of falling yields on the agricultural sector and poor and rural households in Tunisia are significantly negative. Climate change is projected to reduce farm incomes by 2–7 percent annually, to 2030. Rural nonfarm and urban households as net consumers of food will be most affected by the rising global food prices as a result of climate change (chapter 3). The economic and social consequences of climate change include harvest losses, the abandonment of certain crops, increased food insecurity, decreased tourism revenues, and increased water scarcity. All Tunisians will be impacted but particularly affected will be those in already-vulnerable regions highly dependent on agriculture and already suffering high unemployment in the south and center of the country (See chapter 4). The purpose of this chapter is to look at the policy options to address these anticipated climate changes and their associated impacts.

Policy Responses to Climate Change Impacts in Tunisia

Increasing climate resilience in Tunisia will require a diverse set of policy actions aimed at different time horizons and at different actors including all levels of government, the private sector, civil society, and individual households. The policy responses will have to take into account the implications of the Revolution of January 14, 2011, on the systems of government and policy priorities.

The Tunisian Revolution that catalyzed the Arab Spring was driven at least in part by perceptions of regional imbalances and marginalization of inland areas by national development policies. Unemployment and a lack of social protection particularly for young people contributed to the uprisings. Prior to the revolution, economic growth had slowed from around 5 percent in 2006 to less than 4 percent in 2010 and less than 2 percent in 2011 in part due to the 2008 Financial Crises and following turmoil in Europe, which absorbs approximately 70 percent of Tunisian exports and which contributes significantly to tourism revenue. The result has been a rise of unemployment from just over 12 percent in 2006 to 18.9 percent in 2011. The financial sector in Tunisia has been further impacted by the S&P downgrading of Tunisia's credit rating in May 2012.

Responses to climate change will have to contribute to economic growth and employment. Low carbon, climate resilient development must be articulated as a policy priority and demonstrated to be an opportunity for green growth. In addition, climate change responses will need to take into account the greater autonomy of regions in the management of their territories, which will involve greater decentralization of decision making and resources to local governments and communities. Where possible, such strategies must also have a strong territorial dimension and also consider existing imbalances in social protection, employment, and resources, particularly for the poor and vulnerable who are most likely to be severely impacted by climate change. Climate change strategy development can also represent a means to build capacity in public financial management, interagency coordination and planning systems at the national and regional government levels.

Tunisia is already ahead of many Arab countries in developing policy responses to climate change. Tunisia was the first country in North Africa to indicate its support to international climate change processes. It ratified the United Nations Framework Convention on Climate Change (UNFCCC) in 1992 and joined the Kyoto Protocol on January 22, 2003. Tunisia has submitted two national communications to the UNFCCC, the first in 2001 and the second in 2011 and also established a Designated National Authority (DNA) in 2005. The country has developed a number of national adaptation strategies as well as sectoral strategies, including the Strategy on the Adaptation of Agriculture and Ecosystems to Climate Change (January 2007), Strategy on the Adaptation of the Coastal Zones to Climate Change (February 2008), and Strategy on the Adaptation of the Public Health Sector to Climate Change (2010). In addition, in May 2010, Tunisia provided a list of appropriate mitigation actions at the national level to the UNFCCC. This was followed in October 2012 by a document identifying the key areas (transport, construction industry, waste, and so on) where mitigation actions could be undertaken. Tunisia is currently in the process of completing a National Climate Change Strategy, building on and updating previous work.

To be effective, Tunisia's climate change policies need to be based on good knowledge and a thorough review of the current climate, as well as its future and associated uncertainties (see chapter 2); incorporate specific sectoral strategies to address the most vulnerable sectors such as agriculture (see chapter 3); be based on consultation and the expressed needs of the people, particularly the most vulnerable (see chapter 4); and be supported by strong leadership and accompanied by strong domestic policies, legislation, and action plans. It is also essential to mainstream climate change into existing public financial management systems and national policies, plans, and programs. Most of the actions aimed at increasing climate resilience will also have broader local development benefits, by, for example, contributing to job creation, improved environmental governance, and facilitating social inclusion and sustainable growth.

This chapter provides concrete guidance to policymakers in Tunisia on ways to move forward on the climate change agenda at the country level. The chapter does this in two ways. First, it re-introduces the framework for Action on Climate Change Adaptation (Adaptation Pyramid) from chapter 1 of this report. Second, it elaborates five key areas for prioritized policy making including (1) providing reliable and accessible data; (2) providing human and technical resources and services; (3) providing social protection for the poor and most vulnerable; (4) developing a supportive policy and institutional framework; and (5) building capacity to generate and manage revenue as well as analyze financial needs and opportunities associated with climate change adaptation. The policy recommendations within this chapter were developed jointly with the Government of Tunisia (GoT), including broad consultation with national government ministries, research institutions, and local government. The recommendations are based on existing efforts in Tunisia and on the analysis in chapters 1–4. Finally, the chapter presents a policy matrix summarizing key policy recommendations put forward in each of the chapters.

Climate Change Adaptation Should Be an Integrated Part of Public Sector Management for Sustainable Development

Climate change is an essential factor to consider in national and regional planning and public decision making. There are multiple approaches to the effective integration of climate change into policies, plans, and programs in Tunisia. In light of uncertainty climate change mainstreaming should be iterative, flexible, and based on adaptive management. This section elaborates in broad terms simple steps that can be taken by the GoT to establish comprehensive climate change policies, plans, and programs. By following the simple steps elaborated in the Framework for Action below the GoT, in partnership with civil society and the private sector, can ensure that policies, strategies, and action plans build resilience to a changing climate and where possible, promote low carbon growth.

Framework for Action on Climate Change Adaptation
Adaptation is a long-term, dynamic, and iterative process that will take place over decades. Decisions will need to be made despite uncertainty about how

both society and the climate will change and adaptation strategies and measures will need to be revised as new information becomes available. Many standard decision-making methodologies are inappropriate, and alternative, more robust methods for selecting priorities within an adaptive management framework will be more effective.

Elements in an adaptive management model include (1) management objectives that are regularly revisited, and accordingly revised; (2) a model or models of the system being managed; (3) a range of management choices; (4) the monitoring and evaluation of outcomes; (5) a mechanism for incorporating learning into future decisions; and (6) a collaborative structure for stakeholder participation and learning.

In addition, and complementary to an adaptive management approach, Tunisian official could consider five enabling conditions to support the successful integration of climate change adaptation into development processes as articulated by the OECD (2009):

1. A broad and sustained engagement with, and participation of, stakeholders, such as government bodies and institutions, communities, civil society and private sector
2. A participatory approach with legitimate decision-making agents
3. An awareness-raising program on climate change for households, civil society organizations, opinion leaders, and educators
4. Information gathering to inform both national- and local-level adaptation decisions.
5. Response processes to short- and long-term climatic shocks.

These conditions help to ensure that multiple perspectives are brought into the policy decision-making process and therefore help to ensure that policy solutions, that are tried, are robust and in line with an inclusive management approach.

The Adaptation Pyramid introduced in chapter 1, and referenced in chapters 2–4, provides a framework to pursue this adaptive management approach and is designed to assist stakeholders in Tunisia to integrate climate risks and opportunities into development activities (see figure 5.1). It also highlights in particular the importance of leadership, without which adaptation efforts are unlikely to achieve what is necessary to minimize the impacts of climate change.

The base of the Pyramid represents the four iterative steps that form the foundation for sound climate change adaptation decision making:

1. Assess climate risks, impacts, and opportunities for climate action.
2. Prioritize policy and project options.
3. Implement responses in sectors and regions.
4. Monitor and evaluate and subsequently re-assess climate risks, impacts and opportunities.

Figure 5.1 Framework for Action on Climate Change Adaptation: Adaptation Pyramid

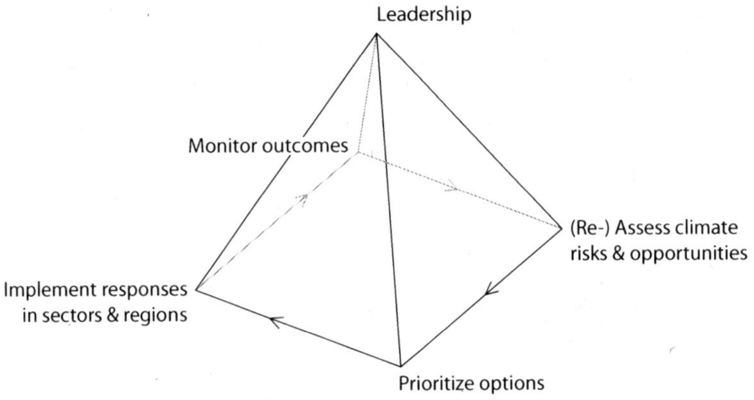

Source: Verner 2012.

The arrows on the four sides of the pyramid highlight the iterative nature of adaptation decision making. Adaptation is a continuous process that takes place over time and adaptation activities will be subject to revisions as new information becomes available. The arrows on the four sides of the pyramid highlight the iterative nature of adaptation decision making. To this we add a fifth apex: that of leadership, without which adaptation efforts are unlikely to achieve what is necessary to minimize climate change impacts. The base of the pyramid is elaborated in the next sections.

Assess climate risk impacts and opportunities: In this first step, a wide range of analyses could be used.[1] All of these rely on access to climate and socioeconomic data to provide information on climate change impacts, including on vulnerable groups, regions, and sectors. To help understand risks and impacts, data are needed on current climate variability and change as well as projections and uncertainty about the future climate. Similarly, information on past adaptation actions and on coping strategies needs to be gathered and evaluated in light of the changing climate.

Prioritize options: The second step is to identify and prioritize adaptation options within the context of national, regional, and local priorities and goals, and in particular financial and capacity constraints. Expectations of climate change make it more important to consider longer-term consequences of decisions, as short-term responses may miss more efficient adaptation options or even lead to maladaptive outcomes, for example the further development of highly vulnerable locations. The prioritization of adaptation responses requires an understanding of the linkages between projected climate impacts, associated socioeconomic impacts, and adaptation responses. One technique for prioritization could include landscape mapping. For example, it may be possible to map where increasing rainfall or aridity will impact current cropping and land use systems, and to similarly map where options exist to move to more drought-resistant crop and forage systems, when increasing aridity makes traditional options no longer viable. Another possible approach to prioritizing options is Robust Decision Making, which seeks to identify choices that provide acceptable outcomes under many feasible scenarios of the future.

Implement responses in sectors and regions: Adaptive responses will often be somewhat at odds with immediate, local priorities, and thus the third step of implementing the agreed responses needs cooperation and understanding at national, sectoral, and regional/local levels (often jointly). At the national level, adaptation needs to be integrated into national policies, plans, and programs and financial management systems. National governments will also have to put in place systems to support local actors, governments, and sectors.

Monitor outcomes: The fourth step is to monitor outcomes to ensure that adaptation-related strategies and activities have the intended adaptation outcomes and benefits. Comprehensive qualitative and quantitative indicators can help project proponents recognize strengths and weaknesses of various initiatives and adjust activities to best meet current and future needs. The monitoring framework should explicitly consider the effects of future climate change, particularly for projects with a long-time horizon.

This is an iterative process: The next step will thus be to reassess activities taking into account new and available information, for example, about future climate change or the effectiveness of previously applied solutions.

The majority of this report has focused on the bottom right-hand side of the pyramid: the assessment of climate change risks and opportunities and the prioritization of adaptation responses. Further guidance will likely be needed to move from prioritized options to effective implementation and monitoring.

Leadership is Central for Successful Adaptation

Effective climate change adaptation in Tunisia will not occur without strong leadership. International experience shows that the lead needs to be taken at the national level by a prominent ministry or senior government champion. In Tunisia, the support of the Prime Minister, as well as other key ministers such as the Minister of Planning and Economy, or the Minister of Environment, will be critical to the effective prioritization and implementation of adaptation actions. This champion will also require the support of a strong team comprised of representatives of relevant ministries, local government, local institutions, civil society organizations, the private sector, academia, and ideally from opposition parties in order to ensure continuity, as governments may change. Clearly, this should be adapted to the context of Tunisia and its circumstances. Leaders are also needed in other levels of government and within civil society and private sector organizations. Leaders from all sectors need support through information access, education opportunities and must be treated as legitimate agents in the decision-making processes. While this report recognizes and emphasizes the importance of leadership to successful climate change action, it does not assess in detail possible champions for climate change adaptation in Tunisia, which can best be determined by the country.

Policy Options Are Available to Support Climate Change Adaptation

This report mainly focuses on assessing climate risk impacts and opportunities and establishes a framework for adaptation decision making. The remaining sections of this chapter focus on the range of policy interventions that are needed in order to increase climate resilience. An understanding of the range of policy options available will enable policy makers to move to the next step in the pyramid—namely policy prioritization. The policy options addressed in this chapter are not mutually exclusive. They aim to show a variety of complementary actions that contribute to five main thematic objectives:

1. Facilitate the development of publicly accessible and reliable information and analyses related to adaptation.
2. Support the development of human, technical, and other resources and services to support adaptation.
3. Provide social protection and other measures to ensure that the poor and the most vulnerable are climate resilient.
4. Develop a supporting policy and institutional framework for adaptation.
5. Build capacity to generate and manage revenue and to analyze financial needs and opportunities associated with adaptation.

The next sections build on chapters 1–4 to provide further guidance on particular policy, plan, and program options available to Tunisia in each of the five thematic areas highlighted above. Table 5.1 provides additional details on the implementation of these policy options in the key economic sectors and areas addressed in this report. Further processes will be required within Tunisia at the national and regional levels to prioritize possible adaptation actions and move toward implementation.

Facilitate the Development of Publicly Accessible and Reliable Information and Analyses Related to Adaptation

Improve Access to Climatological data

Access to quality weather and climate data is essential for policy makers. Without reliable data on temperature and precipitation levels, it is difficult to assess the current climate and make reliable weather forecasts and climate predictions that will allow for the design of effective policies, the implementation of early warning systems, and adaptation within key sectors upon which the economy depends.

Tunisia has regular climate data collection with a high concentration of weather stations in the northeast of the country as well as in Djerba and other

Table 5.1 Data Needs for Effective Adaptation Decision Making

Sector	Types of data needed	Key challenges	Policy options
Climatological	– Temperature – Precipitation – Air pressure – Humidity – Wind – Radiation	– Data often under the governance of the Ministry of Defense, etc., limiting data. – Insufficient data. – Not linked to impact analysis.	– Make climate data available, by the civil authority in charge. – Data rescue and expansion of the number of weather stations. – Ensure data are readily available to policy makers and researchers for analysis.
Food security	– Production levels/yields for indicator crops. – Main food supply chains, models of how they operate, and how they will be affected by climate change. – Imports and exports of key crops as well as food storage. – Operation of safety nets.	– Data to advise on reducing vulnerability. – Cross-sectoral issues and sole mandate of Ministries of Agriculture.	– Identify national and regional partners for data collection and dissemination. – Link findings to early warning systems.
Gender	– Disaggregated by sex, age, and location. – Local knowledge and practices, for example, on local water management systems.	– Data is not always disaggregated by sex and age. – When it is, it is not always analyzed from a gender perspective or publicly available.	– Adjust data collection systems and sets to include information on time-use and division of labor. – Complement with qualitative surveys. – Invest in analysis of existing data.
Health	– Occurrence and magnitude of climate change–related health outcomes, linking those to environmental and meteorological indicators.	– Insufficient data. – Insufficient linking of climate and health data. – Lack of data information systems for monitoring climate and health trends. – Inadequate tracking of vulnerable groups.	– Develop climate sensitive surveillance systems and evaluation techniques for health. – Strengthen health-environment management information systems (HMIS). – Collect and analyze information on groups vulnerable to climate change.
Urban livelihoods	– Geographic location, exposure (river, coast). – Population: absolute and trends. – Risk zones: unstable slopes, low-lying areas, areas of high density. – Governance structure. – Building codes and enforcement. – Economic activities and built environments.	– Data are sparse and often not compiled and analyzed in a holistic way.	– Systematically collect urban data and link with climate change. – Make information available to all including to local level authorities.
Water	– Water availability, salinity, and quality. – River runoff. – Groundwater levels. – Current and future water consumption. – Impacts of various policy measures on water supply and demand.	– Limited capacity to monitor long-term trends in hydrometeorological data and regional climate change modeling capacity. – Limited understanding of the impacts of policy responses on human behaviors.	– Promote regional cooperation and sharing of data and good practices in data collection and dissemination as well as in long-term monitoring, regional water modeling, and economic and policy analysis.

highly developed regions. There are however, low concentrations of weather stations in the southwest and interior of the country. Climate data is currently documented using SDCLIM version 1.0. This is based on individual meteorological elements, which facilitate automatic inputs, but there is some difficulty in the system in terms of introducing new elements or parameters. The system has also only been developed in French, which limits some forms of international and national data sharing (Ben Mansour 2011; WMO 2002). In the short and medium term, the collection and monitoring of climate data could be improved by expanding the number of weather stations, and by collaborating with other countries in the region to improve the coverage and comparability of data. This effort should be combined with a push to link climate data to impact analysis by making climate data available to policy makers and researchers. Some efforts in this direction have already started. Tunisia is part of the European Climate Assessment and Dataset (ECA&D) project. This project, which aims to combine collation of daily series of observations at meteorological stations, conduct quality control, analyze extremes, and disseminate both daily data and analysis results, is gradually being extended across the Middle East and North Africa.

As highlighted in chapter 2, water is scarce in Tunisia and likely to become scarcer due to climate change. Information on current and future water availability and quality is therefore critical for designing adaptation responses. This requires information on river runoff, groundwater levels, and water quality including salinity. While information on hydrology is included in the Climdata system used by Tunisia, in many parts of Tunisia coverage of this data is poor and will need to be upgraded. Capacity is also required to monitor and analyze long-term trends in hydrometeorological data, link it with the climate data, and develop downscaled climate change models.

Link Climate Data with Socioeconomic and Typological Data Sets

Data on climate variability and change, as well as water availability and change, will need to be complemented with socioeconomic data, such as population demographics, growth, employment, health, social security, and so on. At present in Tunisia, ad hoc studies have been undertaken (for example public health studies linking climate and health information). There is not, however, a systematic collection and analysis linking socioeconomic and climate data. The data types needed for effective climate policy making include household data, census data, and other economic data such as labor market and production data. In national, sectoral and local data collection, it is important that social and economic information is collected at a disaggregate format to reflect location, gender, age, and socioeconomic status, as these factors greatly affect exposure to and ability to cope with climate risks. Ideally, micro-data series should be continuous so development over time can be tracked closely.

Location, such as urban or rural areas, is also an important determinant in a changing climate. Climate-related impacts and disasters can be extremely damaging in urban areas. It is anticipated that over 70 percent of the total population of Tunisia will live in urban areas by 2015. Most of these urban areas are on the coast. For effective adaptation planning, data should be linked, such as for example geographical location (GIS) and exposure (for example, proximity to rivers or coasts), as well as information on current and anticipated future population size, and distribution and physical expansion of the urban area. Certain zones within urban areas such as unstable slopes, low-lying areas, or high-density areas will be particularly vulnerable and should be carefully mapped. It will also be important to have information on the potential ability of an urban area to respond and cope with extreme weather events, which will depend on governance structures, building codes and their enforcement, wealth, economic activities, and built environments.

In rural areas, it is important to collect data on changes in agricultural production levels/yields for indicator crops, that is examples from forage, vegetables, grains, and livestock categories.

It is important to ensure that socioeconomic data and policies are then linked with climate change data and policies. For example, with the support of the Social Affairs Ministry, the Internatioal Labour Organization and the Belgian Government, the Tunisian General Labour Union (UGTT) and the Tunisian Industry, Trade and Handicrafts Union (UTICA), aim to establish a diagnosis of the socioeconomic situation in Tunisia and sketch out axes of a social contract for the 2012–20 period. It will be important to consider climate vulnerability as well as climate-related opportunities in the development of these socioeconomic scenarios.

Other data needs identified in this report are presented in table 5.1.

Promote Awareness Raising

Collecting data and analyzing it is an important step, but equally important are the mechanisms for disseminating these data and making them available, so that awareness can be raised and people begin to shift behavior and act or adapt as a result of the information received. In several governorates visited for this report, individuals reported a growing awareness of a changing climate. There was, however, a limited understanding of the linkages between this knowledge and effective climate change adaptation, how this related to the national or international policy dialogue, and how support, for example in developing alternative livelihoods or improved agriculture for drought conditions, could be accessed.

Education has been a priority of the GoT and the country's education system has ranked as one of the best in the Middle East and North Africa in the Human Development Index as well as in Organisation for Economic Co-operation and Development (OECD) and United Nations Development Programme (UNDP) studies. Tunisia has even been one of the few non-OECD countries to participate in the OECD Program for International Student Assessments. Environmental issues have been incorporated into school curriculums through the Sustainable Schools Program.[2] This is a program administered by the Ministry of Environment aimed at environmental education as well as the creation of environment clubs, installation of renewable energy technology in schools, and other environmental conscious activities such as the establishment of eco-gardens. Cimate change could usefully be incorporated into this program as well as into core school curriculums.

The Internet and new media are increasingly popular and accessible even in very remote and poorer regions of Tunisia with many rural schools having computers for general use. The

GoT could use new interest and access to new media to initiate relatively inexpensive climate-related awareness campaigns. For example, awareness campaigns about climate-induced water scarcity may reduce household water consumption, an adaptive outcome. Children with relatively greater literacy in new media and with a tendency to be better educated can play an important role in connecting the traditional knowledge of weather and climate systems of their parents with policy recommendations related to adaptation such as water conservation. The language and methods used to facilitate understanding will need to be appropriate to each locality. In addition, visualizations can enhance understanding. For example, cities, local governments, and universities can employ GIS technology to develop visualizations (often a map) of vulnerabilities and risks.

Businesses and the private sector in Tunisia will also have access to networks important to awareness raising. Therefore, the GoT may wish to leverage private sector finance and interest in climate change awareness campaigns.

These awareness campaigns also need to consider factors constraining women's access to information and target campaigns accordingly. In this regard, female community leaders have a critical role to play.

Provide Human and Technical Resources and Services to Support Adaptation

Specialized human and technical resources are required to analyze, identify, and implement adaptive responses. Human and technical resources can be developed through education and training, research and development, and technical improvements.

There is a need to develop new skills, knowledge, and expertise related to climate change adaptation. This can be achieved through building and/or expanding training, including graduate programs related to climate change and key sectors such as agriculture, health, or water. Moreover, specific training on climatology is important, as a wealth of satellite information is available, but is not being accessed and used effectively by local institutions.

Training can also be targeted to particular areas that are subject to high climate risk. In particularly high risk urban or rural areas, it may be useful to have specific training initially for local governments and planning and emergency management teams on hydro-meteorological disaster management, such as how to respond to floods, landslides, drought, or heat waves, and then subsequent training of the local population by local government and planning and emergency management teams to strengthen responses to these high-risk situations.

A number of education institutes and new environmental education programs in primary and secondary schools already include some climate training. At a graduate level, the University of Tunis offers a range of courses in climatology and climate change. The National Institute of Agronomy of Tunisia includes training in agro-climate linkages and a range of climate-related staff exchanges have been organized between Tunisia and for example the Arid Land Research Centre, Tottori University, University of Lille, University of Oxford, and other institutions. In addition, there have been a range of international conferences on climate change in Tunisia, ranging from the International Conference on Climate Change and Tourism in Djerba, Tunisia (2003), to numerous climate-related workshops hosted by the African Development Bank and GIZ, as well as a World Bank workshop on "Adaptation to a Changing Climate in the Arab Countries" (2012). Tunisia has also sent several government officials and climate change negotiators for climate change training organized by the German Agency for International Cooperation (GIZ), OECD, and the UNFCCC. More could be done to integrate climate change policy into university courses and to create linkages between the technical courses on climatology and general policy making. In addition, further opportunities for staff and student exchanges could be explored.

Adaptation responses can also include specialized training programs for professionals engaged in particular sectors. For example, training for water utility employees to enhance water demand management through market based instruments, including water pricing and metering of water usage, would be a useful adaptation response, and would also result in broader development co-benefits, since the country is already water scarce. In many regions of Tunisia, rural areas are currently experiencing an out-migration of men due to an inability to cope with climate change impacts. For women to in turn cope with the impacts of out-migration and be involved in household decision making, including decisions related to climate change adaptation, special training is required for them in community and political participation skills, business development, as well as general literacy and education and extension services. While Tunisia tends to rank well in the Davos World Competitiveness Report,[3] it has consistently ranked among the lowest in terms of female participation in the workforce ranking 124th out of 133 countries in 2009/10.

Education and training to enable livelihood diversification is also important. A Ministry of Vocational Training and Employment was set up in 1990 but was abolished in 2002. Since then, vocational training has been under the authority of the Ministry of Education. Starting in 1993, the GoT instituted vocational-preparation internships allowing apprentices to obtain vocational training in return for lower wages. In 1997, in the broader context of upgrading the Tunisian economy, the GoT launched a program to raise the standards of training and employment (MANFORME), a strategy of human-resources appreciation focusing on three goals: a demand-induced approach to respond to the production sector; a flexible training capacity to respond to the specificity and evolution of qualification needs; and an encouragement of autonomy in the financing and management of training centres. In other words, all planned actions were to be based on meeting the demand in skills. Since the 1990s, efforts have been made to involve the private sector in training. However, due to the delays in implementation and the lack of financing, many reforms are still in their initial phase. At the end of January 2008, the Chamber of Deputies adopted a government bill on vocational training. The measures aim to increase the intake capacity of the training centres, to reinforce option-based training under partnerships with enterprises and to promote the efficiency of "sandwich-training" schemes. In addition, bridges have been set up between general education and technical learning so students can transfer from one to the other. These efforts need to be further enhanced and targeted towards the needs of a low carbon, climate resilient economy.

Encourage and Support Risk Management Strategies

It is crucial to build a rich and functioning network for risk mitigation that includes social and extension services linking farmers to agricultural research as well as linking vulnerable populations to markets and policy makers. A network of communication and extension services is crucial in order to reach out to farmers, or the agricultural community as a whole. Such a comprehensive system ensures the dissemination of relevant information, as well as techniques and cultivars, and guarantees that national policies are implemented down to the individual unit: the farmer. Furthermore, such a network also provides a strong link back from the farmer to scientists and policy makers for the collection of information relevant for technological advance and policy making. Disaster risk management strategies, such as index-based weather insurance schemes can be a powerful tool to mitigate the risks facing small farmers' livelihoods due to weather variability and consequent crop loss. There are many advantages of simple weather security schemes. Insurance would be provided through groups to reduce transaction costs for the insurance companies (Martins-Filho et al. 2010). Groups would increase coverage on weather variability to small farmers which translates into less livelihood disturbances and risk. In order to successfully operate, a relevant weather index has to be in place. In addition, given the reliance on group insurance structures for these schemes, strong farmer extension channels must also be in place for product and information dissemination.

Encourage Research and Development

Research and development is critical to enhance understanding of current and future climate change impacts and the development of new and appropriate technological responses. Key research often excludes climate related factors. There is a need to work with existing academic institutions or research institutes to strengthen research in areas at the interface of climate change and key issues such as agriculture, gender, health, and urban and rural livelihoods and to develop new specialized institutes or centers of excellence (see box 5.1).

Box 5.1

Research and Development in Tunisia's Agricultural Sector

Investment in agricultural research and development is important for Tunisia, particularly in the breeding of climate-proof crops, as rainfed crop yields are hit especially hard by climate change impacts. Scientific advancement for breeding climate-proof varieties will therefore be key for the future of agriculture in Tunisia. Farmers also have different on-farm management techniques available to offset climate change impacts, which may include: shifting the planting date, switching crop varieties, switching crops, expanding the area of production, and/or increasing irrigation coverage (Burke and Lobell 2010), and improving irrigation

(box continues on next page)

Box 5.1 Research and Development in Tunisia's Agricultural Sector *(continued)*

efficiency. Furthermore, research and development in agriculture would also include changes in crop practices (such as optimum sowing dates, choice of cultivars, and planned plant density (Hainoun 2008)), re-evaluating and redesigning irrigation, and water harvesting practices to sustain a healthy agricultural sector. In trying to address climate change, it is essential to distinguish between short-term measures and long-term measures.

Source: Based on Burke and Lobell 2010; Hainoun 2008.

Enhance Technological Resources

Shifting to new technologies or using existing technologies more effectively could also be a mechanism to build resilience to climate change impacts. See table 5.2 for a list of low carbon and climate resilient technologies that could be applied in the agricultural sector in Tunisia. The GoT has an important role to play in facilitating promotion of and access to technologies that help people to adapt to climate risks. Recent agreements with India to promote cooperation in biotechnology research offer some prospects for future growth, and additional agreements related to water and energy technology within the region and beyond would be useful. In addition, as illustrated in box 5.2 efforts are under way to address water shortages through desalinization technology.

The GoT also has a critical role to play in promoting technology for enhanced water supply and decreased water demand. This is best accomplished through a combination of policy reforms that change incentives for private investment in new technology for greater climate resilience and address key market failures, combined with public financial interventions or investments. In terms of water supply, new or improved technology could be used to reduce water network leakage from pipe systems or to improve storage and conveyance capacity.

Equally important, water demand could be reduced through, for example, drip irrigation or better metering. Desalination, an important current technology for enhancing water supply, can result in significant local environmental impacts, including pollution to both air and water, thus affecting human health, marine environments, and economic activity, such as local fisheries. Developing and shifting to new desalination technologies (such as those described in box 5.2), which reduce both air pollution and brine discharge, could help to minimize impacts and facilitate adaptation responses.

These technologies can be locally derived through processes or research and development or made accessible through technology transfer. There are, however, often barriers to appropriate technology development, transfer, and use that must be overcome. These relate to inadequate information and decision support tools used to quantify and qualify the merits of various low carbon or climate resilient technologies and related investments, as well as limited local human and technical skills and the need for appropriate training. These can be overcome with voluntary and regulatory approaches to enhance coordination

Box 5.2

Ksar Ghilène (Tunisia) Stand-Alone/Autonomous Brackish Desalination Plant with Reverse Osmosis (RO) and Photo Voltaic (PV)

Project objectives	Technical features	Achievements	Challenges and new pilot projects
Desalinate brackish water from traditional oasis with PV solar energy and RO in an isolated traditional community.	**Well** Brackish water comes from an artesian well inside the oasis, 2 km away from building project. Aproximately 3 m³/h is pumped with natural pressure (6 bars) with variable temperature (37°C in summer and 13°C in winter). Salinity of brackish water: 5,700 TDS or 3,500 mg/l	Average annual water production: 7.5 m³/day.	Disseminate results, information and raise public awareness.
Provide potable water to a village of 300 habitants, 150 km away from closest grid network and 60 km away from the next well (tank mobile transport).		Average annual energy consumption of 15 kWh (total recovery of energy is about 70 percent).	Promote interest of regional R&D institutions, promoting patent and technology transfers (public and private partnerships).
Support main subsistence activities: agriculture, cattle raising, and sustainable tourism.	**RO desalination plant** Brackish water is stored in a 30 m³ tank with chlorine dosing. Pretreatment system is composed of a sand filter, carbon filters and a pretreatment with a cartridge-type filter. The RO module has a 2.1 m³/h capacity, 50 m³/day (<300 μS/cm conductivity). The system is completed with a 25 m³ treated water tank for storage and a brine discharge (0.9 m³/h).	Street lighting works with photovoltaic streetlamps. Electrified houses by Solar Home Systems.	Promote interest of international development institutions and disseminate results within the international community for adaptation to climate change in MENA region.
Support social infrastructure: school, mosque, national guard office, and a clinic that provides medical services once a week.		Potable water for sanitary uses and clinic operations.	Desalination powered by renewable energy, promoted by the European Union (EU), are projected in Turkey, Morocco, Jordan, and Egypt (Master Plan for the Wide Implementation of Autonomous Desalination Systems, 2007) (www.adu-res.org).
Starting operation: the fountains of the village were opened for the water supply to the population.	**Solar Photovoltaic System-stand-alone Electric Network** 7 PV generators in parallel of up to 10.5 kW peak power of the field of panels. The PV panels provide energy to the isolated electric network: a battery bank (660 Ah C10 of capacity at 120VDC, giving 10kW of nominal power) which controls the battery charge and operates as an inverter to keep AC signal constant at 230V	No use of fossil-fuels in the village. Clear improvement of living conditions (social stability, sanitary and medical services, plus a major population accession).	Pilot ADIRA project in Jordan (Hartha Charitable Society in the north): RO-PV installed and operated in a rural village; specific energy consumption of the unit is 1.9 kWh/m³. (Qiblawey, Banat, and Al-Nasser 2011)
Project developers and financing: Government of the Canary Islands, Instituto Tecnológico de Canarias (ITC) and Agencia de Cooperación Internacional (AECI) from Spain; Agence Nationale de Maîtrise de l'Énergie (ANME) and Commissariat Régional au Développement Agricole de Kébili (CRDA) de Kébili (Tunisia).		Theoretical and practical training for local technicians in charge of supervising the system, its operation, and its maintenance.	
References: www.prodes-project.org and www.itc.org		Building construction taking into account climatologic conditions and sandstorms (semi-buried).	
Current status: operating			

Source: Noemi Padrón based on Manufacturing Company/Responsible Organization: Institut National de Recherche Scientifique et Technique (INRST), Tunisia. http://www.prodes-project.org.

Table 5.2 Climate Resilient and Low Carbon Technology in the Agriculture and Water Sectors in Tunisia

Integrated policy objectives	Main benefits	Available Technology and R&D	Identified barriers
Increase agricultural productivity and efficient water use through technological approaches to adaptation in the agricultural sector	– Reduce water stress (and desalination needs) and groundwater overexploitation, increase productivity and available water. – Reduce vulnerability to rainfall variability. – Increase food security and reduce poverty (food buyers and exports) by reducing food costs. – Improve livestock management and health positive externalities.	– Reduce water losses and improve productivity through drip, sprinkler, and bubbler irrigation systems. – Affordable Micro-Irrigation Systems (AMITs). – Fog harvesting as applied in Yemen (26 small Standard Fog Collectors [SFC]). – Strengthened weather information, covered agriculture, alternative seeds and saline water research.	– Low technological input and low grid expansion in remote areas. – Investment needs for smalls projects. – Cross-subsidized tariff favoring irrigation water. – Need for demand management programs (to drive technological change and promote intersector water transfers). – Needs for water storage. – No consideration of hydro-footprint in trade and water added-value.
Promote investment and encourage research networks to increase wastewater treatment in urban areas for agricultural water use and natural wastewater systems for remote and isolated human settlements	– Increase available water in remote areas and reduce water stress. – Reduce land and water pollution. – Decentralize systems to allow for multiple secondary benefits (artificial wetlands, lagoons and land recovery).	– Long experience in using treated wastewater to irrigate: citrus orchards and olive trees of the Soukra irrigation scheme, 600 hectares. – More than 61 plants collecting 0.24 billion km^3 of wastewater (less than 30 percent is reused to irrigate vineyards, citrus, trees, fodder crops, industrial crops, cereals, and golf courses in Tunis, Hammamet, Sousse, and Monastir. – Secondary levels and farmers pay subsidized prices for irrigation with reused water.	– Water supply driven policy but civil participation and maintenance. – Low sewage networks in remote (and some urban) areas. – Salinity problems. – Low storage capacity.
Promote desalination with renewable-energy desalination systems to improve water shortage in rural and isolated communities **Spread pilot projects of autonomous desalination systems such as Ksar Ghilène**	– Increase potable water from brackish water and polluted sources. – Reduce grid investments. – Simple construction and maintenance. – Abundant solar and wind sources. – Increase energy supply in remote and farming areas (especially in the off-grid south communities).	**Desalination with Solar Energy:** Sea water desalination with Solar Collectors in the **Area of Hzag,** (0.1–0.35 m³/h); Sea water distillation with Mutti-Effect Distillation, Solar Collectors (120 m²), **Beni Khiar,** (1 m³/day); Sea water desalination with Membrane Distillation, compact system or fully solar driven combining Solar Thermal Collectors (6 m²) and Photovoltaic (80 Wp), in **Tunis,** (150 lt/day); Brackish water desalination plant, Solar Collectors (80 m²), Hazeg, Sfax, (400–500 lt/day). **Desalination with Photovoltaic Energy:** Autonomous brackish desalination (RO) plant, driven by Solar PV Energy (10.5 kW PV) in **Ksar Ghilène,** (2.1 m³/h, 15 m³/day); Brackish water desalination plant (Reverse Osmosis) plant, driven by Solar Photovoltaic Energy (590 Wp) in **Hammam Lif** (50 lt/day).	– Low production and expensive components (0.1–20 m³/day and 5–10$/m³. – No local industries and need to provide training. – Scarce cooperation in R&D, local components and patent transfer. – International cooperation projects with scarce continuity and positive externalities. – Subsidized oil and direct subsidies in energy intensive sectors. – Low private sector involvement in RREE projects. – Low participation in CDM projects. – No integration of regional markets.

Source: Noemi Padrón based on Manufacturing Company/Responsible Organization: Institut National de Recherche Scientifique et Technique (INRST), Tunisia. http://www.prodes-project.org.

and information sharing between governments, local people, and the private sector, as well as through revision and clarification of laws related to environmental technology and local training and incentive programs.

Build Climate Resilience of the Poor and Vulnerable Through Social Protection and Other Measures

As highlighted in chapter 4, vulnerability to climate change impacts depends on two things: the scale of climate change impacts, and human resilience as determined by factors such as an individual's age, gender, or health status; a household's asset base and degree of integration with the market economy. Underinvestment in social safety nets, public services such as water supply and wastewater treatment, and housing and infrastructure in vulnerable areas make poor people more vulnerable to a changing climate. Therefore social protection and other measures to ensure that basic needs are met are critical instruments to build resilience to climate change for the poor. Since these same instruments facilitate economic and social inclusion, there are also clear development cobenefits in investing in these measures.

In response to the revolution of January 14th, 2011, the GoT is seeking to reform its social protection system with a study of social vulnerability currently under way. At present, however, climate change has not been considered as a key factor for social vulnerability in this study. In the context of increasing climate change impacts, which particularly affect the poor and vulnerable in many regions of the interior, it will be essential to incorporate climate change into the study.

A good climate change adaptation strategies for the rural and urban poor is good development policy. Even if the severity and frequency of climate change impacts remain constant, these are likely to have increasingly negative socioeconomic consequences as a result of a larger population, through an increasing demand for food and increasing groundwater depletion. Rural farm households, rural nonfarm households, and the urban poor are particularly hard hit, mainly due to the large share of their income they spend on food and the reliance of rural farm households on "on-farm" income for their livelihoods. Social safety nets are critical in order to provide the necessary channels of outreach and support to the poor and vulnerable, both in times of crisis and under more normal conditions In general, social safety nets and long-term development goals should be integrated together with national planning goals and objectives.

Measures to ensure social protection can include insurance schemes, pensions, access to credit, cash transfer programs, and relocation programs. Additionally, it is important to ensure that the poor can meet their basic needs and that there are measures in place to guarantee access to affordable health care and education. At the local level, livelihoods and assets can be protected through various forms of insurance such as life insurance, infrastructure insurance or weather based index insurance (box 5.3).

In Tunisia's rural areas currently suffering from declining agricultural yields and the outmigration of men, social protection is particularly critical for women,

Box 5.3

Weather Based Index Insurance

Index insurance represents an attractive alternative for managing weather and climate risk because it uses a weather index, such as rainfall, to determine payouts. With index insurance contracts, an insurance company does not need to visit the policyholder to determine premiums or assess damages. Instead, if the rainfall recorded by gauges is below an earlier, agreed-upon threshold, the insurance pays out. Such a system significantly lowers transaction costs. Having insurance allows policy holders to apply for bank loans and other types of credit previously unavailable to them. However, if index insurance is to contribute to development at meaningful scales, a number of challenges must be overcome, including enhanced capacity, establishment of enabling institutional, legal, and/or regulatory frameworks, and availability of data. Droughts, floods,, and other extreme events often strip whole communities of their resources and belongings. Index insurance, however, could enable the poor to access financial tools for development and properly prepare for and recover from climate disasters.

Source: Barrett et al. 2007.

the elderly, young people, and children that are left behind. It can take the form of rural pension schemes, or conditional cash transfer programs. Other assistance to enhance productivity can include access to credit and facilitated access to markets for agricultural and other rural products in order to enhance income despite declining agricultural productivity due to climate change impacts.

In urban areas such as Tunis, social services can include the provision of affordable housing away from zones at particularly high risk from climate impacts such as flood zones or drought areas. Additionally, the poor and vulnerable will require additional measures to increase resilience, such as conditional cash transfer programs similar to *Bolsa Família* in Brazil, and an affordable provision of basic services such as energy essential for heating and cooling, water, and public transport.

The poor and most vulnerable are particularly in need of assistance when an extreme weather event or other climate change–related crisis hits. Reducing vulnerability to climate impacts for the poorest should therefore be integrated into emergency planning. This could include the provision of basic needs, including adequate shelter, access to food, water, and clothing.

Develop a Supportive Policy and Institutional Framework for Adaptation

A supportive policy and institutional framework in Tunisia at the national, sectoral, and local levels is essential for effective climate change adaptation decision making. Basic conditions for effective development such as the rule of law, transparency and accountability, participatory decision-making structures and reliable public service delivery that meets international quality standards are conducive to effective development and adaptation action. In addition, climate change adaptation requires new policies and structures and changes to existing policies. These include the finalization of the climate change adaptation strategies already under way, supportive policies at all levels, the mainstreaming of climate change considerations into existing policies as well as the creation of systems for cooperative and coordinated decision making between government departments, different levels of government, and cooperation with civil society, the private sector, and other states. This will require a clear coordinated governance structure and likely enhanced training and support for staff responsible for collaboration and coordination.

National Policies Related to Climate Change

Climate change should be a central component of national policy making in Tunisia. With the new government, the Presidential Program 2009–14 "Together to Meet the Challenges" is undergoing revisions and a new constitution is being introduced. Climate change should be a central element of this new constitution and the revised Presidential Program.

In addition, Tunisia is increasingly showing progress integrating environment and climate change into growth strategies. On May 23, 2012, Tunisia signed up to the OECD Declaration on Green Growth, which represents important opportunities for policy cooperation to achieve low carbon and climate resilient growth. Tunisia has also been engaged in the Green Growth strategy development for Rio+20 which includes elements of both low carbon and climate resilient growth.

Develop a National Adaptation Strategy

Tunisia has already developed a number of national adaptation strategies as well as sectoral strategies such as the strategy on the adaptation of agriculture and ecosystems to climate change (January 2007), the strategy on the adaptation of the coastal zones to climate change (February 2008), and the strategy on the adaptation of the public health sector to climate change (2010).

It has been recognized that in order to create coherence between these various sectoral strategies as well as between adaptation and mitigation strategies, there will be a need for an overarching climate change strategy at the national level. Throughout 2011 and 2012, a process has been under way in cooperation with the GIZ, World Bank, and other donors to develop this National Strategy for Climate Change. The approval of this draft strategy however has been delayed due to the political transition. It will be important for this strategy to be approved before the UNFCCC Conference of Parties in Qatar, in order for Tunisia to make the best use of new opportunities for international leadership and climate change financing in the wake of the agreements in Durban. For example, to make use of the adaptation framework agreed in Cancun and formalized in Durban, Tunisia needs to be able to clearly put forward a climate change strategy and articulate financial, technological, and other needs for adaptation. Tunisia, also has a unique geopolitical position and demographic with strong ties in both the Northern and Southern Mediterranean. It could therefore play an important leadership role in establishing a common climate policy across the Mediterranean. However, such capacity is limited without a national overarching climate change strategy.

Highlight the Importance of Agricultural and Water Considerations

Separate sections of national strategies or stand-alone policies related to agriculture and water are useful given their importance for well-being and income generation, as discussed in earlier chapters of this report.

From a process standpoint, an important mechanism to encourage dialogue and coordinate government response to reduce climate induced risks to food security for example, could be a coordinated interministerial national policy or working group supporting food security and rural livelihood development.

Develop a Comprehensive Food Security Strategy to Prepare Tunisia for Rising and More Volatile Global Food Prices

Given that climate change translates into higher and more volatile global future food prices and that Tunisia is likely to become more food import dependent in the future, a forward looking national food security strategy is urgently needed. Important components of such a strategy include (1) assessing the potential and future role of agriculture for the economy and for food security; (2) determining future allocations of water between agriculture, commercial, and household use; (3) revisiting international trade agreements and domestic food supply; chains; (4) assessing health and policies related to population growth, poverty reduction, and nutrition; and (5) managing food security risk.

Integrate Adaptation Considerations into Existing Policies, Plans, and Programs

Integratation or mainstreaming of climate change into all major policies plans and programs is critical. Climate and economic and social development are interdependent. The way countries manage the economy and political and social institutions have critical impacts on climate risks. Therefore mainstreaming climate change is critical to development planning and policy formulation.

Climate mainstreaming is the process(es) by which climate considerations are brought to the attention of organizations and individuals involved in decision making on the economic, social and physical development of a country (at national, subnational and/or local levels), and the process(es) by which climate is considered in taking those decisions.

The process is based on an analysis of how climate change may impact a policy, plan, or program. This includes an analysis of the extent to which the activity under consideration could be vulnerable to risks arising from climate variability and change; the extent to which climate change risks have been taken into consideration in the course of formulation of the existing policy plan or program; the extent to which the activity could lead to increased vulnerability, leading to "maladaptation" or, conversely, miss important opportunities arising from climate change; and what amendments might be warranted in order to address climate risks and opportunities.

It is particularly important to mainstream climate change adaptation considerations into vulnerable sectors in Tunisia, such as agriculture, health, trade, tourism, and water at all levels. While there are strategies/think pieces related to climate change and agriculture, and climate change and health at the national level, and efforts are under way to develop a national climate change strategy addressing all of these sectors, climate change is not as yet mainstreamed into other national policies in Tunisia (for example, policies related to infrastructure development). Climate change considerations will need to be incorporated into infrastructure development in both rural and urban settings, particularly in areas prone to extreme weather events.

In rural contexts, climate change considerations need to be incorporated into existing land property rights and practices. Rural areas are experiencing climate-related outmigration, particularly of men who have traditional held inheritance and land rights. More equitable rights for women will help reduce both the vulnerability of women remaining in rural areas and their entire household.

The Reform of Tunisia's Energy Strategy Could Offer Important Opportunities for Economic Growth, Increased Climate Resilience, and Climate Change Mitigation

Tunisia already has a strong energy sector. Tunisia's oil and gas resources are modest by international standard (and compared to those of neighboring Algeria and Libya), but they make it possible for the country to limit its import dependency. Tunisia is a net exporter of oil and oil products (although it has to import about half its needs of products for lack of refining capacity). Gas accounts for just below 50 percent of primary energy needs and almost 60 percent of it is domestic. The share of renewable energy is still very limited, but its potential is considerable, given the country's solar and wind potential. Tunisia is almost totally electrified (with an electrification rate over 99 percent) and access to other forms of energy is also very high. Some quality issues remain but are limited (voltage/frequency of electricity, quality of gasoline, and so on). The *Société Tunisienne de l'Electricité et du Gaz* (STEG) is considered a well-managed utility compared to similar companies in the region. Comparisons with such companies indicate unit fuel consumption below that in the other

COMELEC countries; transmission losses and outages are also lower than in the other Maghreb countries. However, STEG is heavily subsidized (directly, and indirectly through gas prices), which constitutes a heavy burden on the national budget.

Tunisia's energy sector also benefits from strong private sector involvement. State-owned enterprises (SOEs) are dominant across the energy sector. However, private sector participation is already the norm in upstream oil and gas (which is usually carried out in conjunction with the relevant SOE) and the distribution of oil products (which are either imported or refined by SOEs). The first independent power producer (IPP), Carthage Power, went online in 2002 and now generates over 20 percent of the country's electricity. Additional IPPs are planned, including a 1,200 MW capacity as part of the ELMED project, which would be partly dedicated to exports.

In addition, Tunisia already has a strong tradition of energy conservation. Tunisia has been a pioneer among developing countries for energy management, having formulated and implemented a policy for rational use of energy and promotion of renewable energy as early as 1985. The 11th Development Plan covered the period 2007–11 and the more recent Tunisian Solar Plan covers the period 2010–16. They both aim at promoting energy efficiency and increasing the share of renewable energy in the national mix. Energy intensity in Tunisia remained stable along the 1990s and started to decrease significantly during the 2000s as illustrated in figure 5.2.

In spite of the strong energy efficiency policy, Tunisia has experienced growing energy imports, growing emissions and growing exposure to international markets, with negative consequences in terms of cost, volatility and climate change. Until the late 1980s, Tunisia was characterized by an energy surplus;

Figure 5.2 Tunisia: Energy Intensity 1990–2010

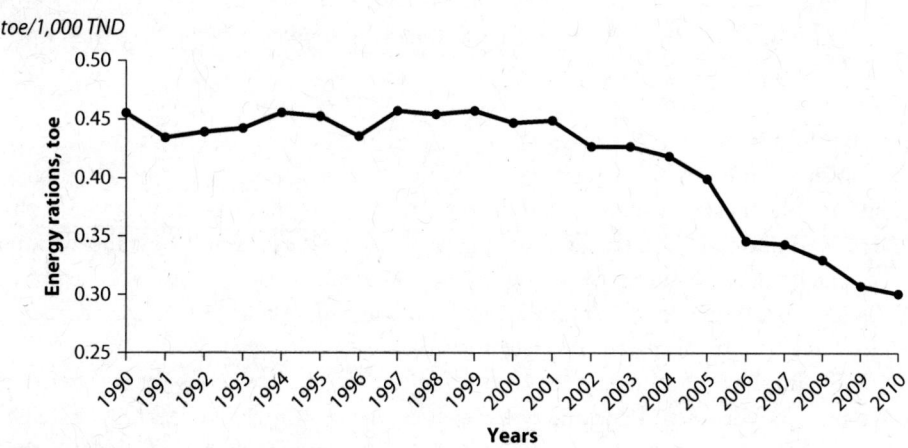

Source: Ministry of Industry and Technology.
Note: toe = tons of oil equivalent.

Figure 5.3 Tunisia: Energy Trade Balance 1990–2010

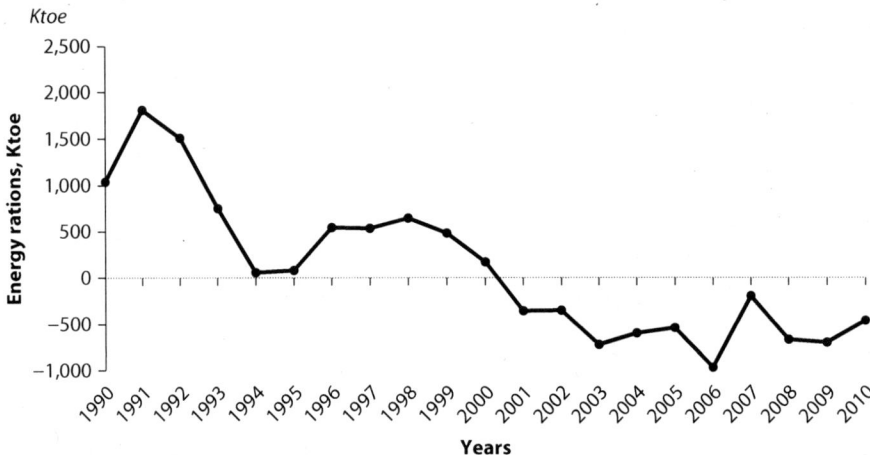

Source: Ministry of Industry and Technology.
Note: Ktoe = thousands of tons of oil equivalent.

this surplus has turned into a deficit and Tunisia became a net energy importer in 2001 (See figure 5.3). The country's primary energy demand is nearly entirely based on oil and natural gas, and its electricity production on natural gas only. Almost 40 percent of Tunisia's gas consumption is currently imported from Algeria and, in order not to become too dependent on its neighbor and on gas imports for power generation, the country is considering various options for the future power plants, including coal, nuclear, and renewable energy.

This energy importation is creating a growing burden on the budget. Energy is subsidized in Tunisia, be it electricity or oil products, which is becoming more and more costly to the country in a context of high international energy prices and scarce fiscal resources (fuel subsidies reached an average 1 percent of gross domestic product (GDP) during the period 2005–10, Fattouh and El-Katiri 2012). It is also a concern in terms of climate change as it favors less than optimal energy consumption.

Toward a New Energy Strategy
Tunisia's strategic objectives for its energy sector are as follows: (1) provide affordable energy in the best conditions for all Tunisian citizens throughout the country, and (2) leverage the energy sector to create jobs and develop regions throughout the country. Job creation and regional development can contribute to alternative livelihood creation in rural areas vulnerable to climate change—therefore contributing to adaptation.

To implement these objectives, Tunisia intends to consolidate existing resources. This means putting less pressure on fossil fuel reserves through sustainable extraction of hydrocarbons, as well as consolidation of the existing energy efficiency policy and development of new, clean, domestic resources. The goal is to build a new energy model that will allow the development of renewable energy resources throughout the country, including provision of components and services. International energy cooperation with North African and European countries will also be developed, with renewable energy as a cornerstone of such cooperation (regional grid management to better overcome intermittency of renewable generation and share capacity reserve, export of green electricity to Europe, and so on).

Four pillars have been defined by the Ministry of Industry and Technology. The World Bank intends to offer full financial and knowledge based assistance for each one of them.

Pillar 1: Resources & Infrastructure

Tunisia needs to optimize its energy mix. The power sector is mostly based on natural gas. It is a fairly clean fuel, but it is increasingly imported, and from only one country, Algeria, which is cause for political and economic concern. Tunisia therefore intends to optimize its energy mix. This involves the following:

- Develop exploration and production of domestic hydrocarbons (conventional and unconventional).
- Continue promotion of energy efficiency, based on existing policy and instruments.
- Develop renewable energy and build on the country's considerable potential in solar energy, wind energy, and biomass.
- Develop pumped hydro storage to support other renewable sources, although Tunisia's hydropower potential is limited, STEG intends to develop a 400–500 MW pumped storage unit, which would be key to optimize utilization of other renewable resources (pumped storage allows to overcome the intrinsic intermittency of wind and solar).
- Explore other options, including coal and nuclear. As fossil fuels are expected to dominate Tunisia's energy mix for a long time, even if the carbon intensive coal option is not selected, carbon capture and storage (CCS) options will also be explored for power plants and major industrial carbon dioxide (CO_2) emitters.

A key element of this pillar is the Tunisian Solar Plan, which was launched in December 2009 for the period 2010–16 and aims at increasing the share of renewable energy and energy efficiency: 40 projects, which will also be part of the Mediterranean Solar Plan, have been identified in solar, wind, biomass, and other sectors, for a total investment amount of €2 billion, 70 percent of which is to be provided by the private sector.

Tunisia proposed four projects in the World Bank/African Development Bank sponsored Middle East and North Africa (MENA) Investment Plan for Concentrated Solar Power (CSP), for which US$ 180 million of concessional funding are earmarked by the Clean Technology Fund (CTF). Those projects include a 50 MW plant under development by STEG, a 50–100 MW private project and two components of ELMED, a joint project between Tunisia and Italy. The CTF would contribute in financing 100 MW of CSP capacity (as part of the 1,200 MW generation capacity planned for ELMED) and part of the planned Tunisia-Italy interconnector. This interconnection with Europe is a key element for the MENA-Europe energy integration and for the viability and sustainability of Tunisia's renewable energy program (thanks to highly profitable potential exports).

Beyond interconnection with neighboring markets, optimizing the energy mix will translate into considerable investment needs for the decades to come throughout the value chain: production, treatment/refining, transportation, distribution, storage, and associated manufacturing and services.

Pillar 2: Employment & Regional Development

Another central feature of Tunisia's new energy strategy is the medium to long-term development of a local equipment industry to contribute to economic growth and job creation throughout the country. The 2011 study initiated by the World Bank on the local manufacturing potential for CSP components in MENA[4] confirmed that such potential exists in most countries of the region, including Tunisia, which is already home to a significant number of suppliers and subcontractors to the European automotive industry. Many of these companies will be able to diversify into green tech-related components and systems, as many of their skills are transferable (metallurgy, mechanical engineering, electrical equipment, and so on).

The government intends to reinforce national capacities through education and professional training, and through better targeting of research and development. Reinforced human capital will allow increasing local integration throughout the clean energy sector. Building on education, training and R&D, and on the strong existing engineering and manufacturing base, Tunisia should be able to build a competitive energy industry, the best potential being in renewable energy and energy efficiency.

Pillar 3: Sector Reorganization

Tunisia's legal framework needs to be adapted to comply with and support the new requirements of the energy sector:

- Update of the energy efficiency framework and incentive system, including norms and standards, subsidies, and financial instruments (subsidized loans to enterprises and individuals, guarantees, and so on)
- Update of the renewable energy regulation, including off-take arrangements with STEG (prices and volumes), third party access to the grid and the supply market, direct export agreements, and so on)
- Assessment of innovative instruments and their suitability to the Tunisian market, including feed-in tariffs, green/white certificates, carbon taxes, and so on.

The energy pricing system needs to be based on the real cost of the energy: Tunisia is planning to work on the progressive removal of subsidies on all fossil fuels and electricity. This should be a key trigger to further promotion of energy efficiency, low carbon growth and climate change mitigation.

In order to favor private sector involvement, the legal framework for independent power generation also needs to be updated, including large -scale supply to the grid, small-scale renewable projects, self-generation, and future export-oriented projects.

Pillar 4: International Integration
International energy integration relates to both infrastructure and markets:

- *Infrastructure*: development of electricity interconnection with Italy (ELMED), regional interconnection of gas and electricity grids within Maghreb countries (including Libya)
- *Markets*: integration into a regional Maghreb power market and with the European Union markets, which will require harmonization of rules and regulations regarding market operation, and will at the same time provide an opportunity to improve the region's security of supply.

Horizontal and Vertical Collaboration Is Critical for Effective Climate Change Policy Making

Within national governments interministerial coordination is critical as adaptation responses often require activities involving multiple ministries and sectors. Interministerial coordination can be achieved through interministerial committees, for example having climate change focal points in all relevant ministries. The private sector and academic and research institutes can also be integrated into these committees as technical advisors.

Coordination between different levels of government is also essential, as climate change adaptation policies will ultimately be implemented by sectoral authorities, local officials, and citizens themselves. For example, to create connections between the Ministry of Agriculture and local farmers on climate risks and adaptation options, it may be possible to use existing farmers' associations that link directly to agricultural research/extension services. This would ensure clear flows of knowledge to all areas from top-down and bottom-up, relying on existing institutional mechanisms that are already in place.

Regional and International Collaboration Is Also Essential for Building Climate Resilience

The heterogeneity of the Arab countries provides multiple opportunities for beneficial climate related regional collaboration between Tunisia and its Arab neighbors. Tunisia will be best equipped to address climate change if it continues to have strong collaboration with other Arab countries as well as the European Union (EU) on issues such as climate-related data sharing, crisis responses, and the management of, for example, disease outbreaks, migration, shared water resources, as well as strong trade relationships to address food security, and to develop economic competitiveness and employment. The World Bank's Doing Business 2010 report ranked Tunisia among the 10 best ranked countries in the Arab World due to reform in the tax system, welfare system and trading including Tunisia's one-stop e-window for trade (Tunisia Trade Net) intended to simply procedures for trading across borders (World Bank 2009). Tunisia needs to build on this strong record and rebuild trade relationships that have been damaged by the political instability postrevolution.

Where knowledge, skills, or technology are lacking in Tunisia, they often exist in other countries. Therefore collaboration with other countries and regions for example in health or management of shared water resources could be particularly valuable. Tunisia could also consider establishing foundations or centers of excellence for example in climatology or in "climate change & public health"-related disciplines. Knowledge sharing can be promoted on a regional scale, for example through staff exchanges, and through enhanced regional and international cooperation such as through the creation of an Arab knowledge network on climate change adaptation.

Opportunities to engage with existing or new initiatives within international bodies are key for improved policy making in Tunisia. For example, the World Meterological Organization (WMO) is promoting a new large-scale initiative on climate services that could benefit from enhanced participation from Arab States including Tunisia.

Build Capacity to Generate and Manage Revenue and to Analyze Financial Needs and Opportunities

Financial resources are essential for development and to effectively adapt to climate change. Tunisia will need to invest in capacity to generate and manage climate change related resources from both domestic and international sources and to analyzing financial needs related to climate change.

Ministries will need to include climate change in sectoral budget estimates and allocations related to agriculture, energy, health, tourism, transport, and water. Moreover, current and future climate change impacts need to be taken into account in planning and costing investments, particularly long-term investments. Financial resources for climate change will need to come from domestic and international sources. National public expenditure reviews could be one tool to highlight current expenditures and hence better understand how these relate to budget estimates for climate proofing infrastructure. This information, in turn, could help the GoT to understand what levels of additional revenues are needed to make up the shortfall and to identify new revenue opportunities, such as payments for ecosystem services, removal of subsidies, or innovative tax mechanisms.

At the local level, access to financial services can play a critical role in helping the poor widen their economic opportunities, increase their asset base and diminish their vulnerability to external shocks including climate change. In rural areas, simple financial services such as credit and saving, can directly affect small producers' productivity, asset formation, income and food security. Payment for Ecosystem Services (PES) has significant potential to enhance rural livelihoods and agricultural yields, maintain and enhance ecosystem services, such as watersheds and biodiversity, and develop long-term partnerships with the private sector. PES can contribute to disaster risk reduction, with the revenues generated serving as financial buffers for communities to climate-induced shocks.

Provided with financing opportunities and incentives, smallholders and rural communities can invest in preventing natural disasters by maintaining sand dunes, conserving wetlands and foresting slopes as cost-effective measures, while at the same time protecting their own assets and livelihoods. Dependable revenue streams would allow them to invest in their crops and land, thus strengthening their businesses.

Funds can be accessed from international sources. Funding channels are multiplying. Particularly important for adaptation are the UNFCCC Adaptation Fund, the UNFCCC and GEF-administered Least Developed Country Fund (LDCF) and Special Climate Change Fund (SCCF), the Pilot Program for Climate Resilience (PPCR, under the Climate Investment Fund managed by the Multilateral Development Banks) and the many bilateral funding arrangements. The OECD has estimated that in 2010 about US\$3.5 billion was provided by OECD members to support adaptation activities with another US\$6 billion for "adaptation related" activities. For a summary of options available, see www.climatefinanceoptions.org. The World Bank is already providing funding to countries through ongoing technical assistance and lending operations and the Climate Investment Fund (specifically though the Strategic Climate Fund's Pilot Program for Climate Resilience).

Improving Resilience and Adaptability of Rural Communities

Ultimately climate change impacts in Tunisia will be felt at the local level particularly in the Central and Southern Regions. In these regions climate change impacts are multifaceted and complex. Food production systems and the agro-ecological conditions sustaining local livelihoods in these regions are severely stressed. The combination of historical overexploitation of natural resources, primarily water and soil, and the climate-related stress on living and production conditions have had negative impacts on rural income generation and employment; on food security, both at the household and national levels; as well as on natural resources, particularly water, and it is here that effective policy responses will be most critical. Furthermore, policies of decentralization following the revolution of January 14, 2011, are increasing the policy making power of these regions. There is therefore a strong need to translate adaptation policy recommendations to the level of the region or governorate with specific recommendations for local governorate actors.

A sound working relationship between communities and key local government actors, particularly agricultural extension service agents, is paramount to build adaptation responses that respond to specific community vulnerabilities. There is also a strong link between regional development and territorial planning, particularly in relation to access to basic infrastructure. There is also a need to support local actors, governments and sectors to implement and adaptation activities. This will require an ability to cope with the multiple, and often related, uncertainties related to planning despite the variability and difficulties in predicting rainfall, or extreme events at the level of individual governorates. The Adaptation Pyramid highlighted above provides an iterative approach to decision making that integrates uncertainty.

It will be particularly important to diversify local livelihood and production options. Maintaining people on the land with viable income generating opportunities is especially important in terms of preserving social structures, by reducing migration and absenteeism. Experience and knowledge of the land represents in itself a tool for adaptation. The expansion and continuity of traditional (for example, dates and olives) and alternative (for example, cactus, native fodder species) agricultural products will need to be balanced against water availability and the drought-resistance capacity of these species. In terms of potential adaptation responses an integrated management approach to agro-ecosystems and associated natural resources, particularly water and soils will be required. There will also be a need for diversification of production systems, providing communities with alternative income sources. Boxes 4.8 and 4.9 in chapter 4 provide two examples of successfully integrated agro-ecosystem management and diversification of agricultural revenue sources for rangelands and oases.

In conclusion, this report is intended as a resource for Tunisian policy makers to begin to assess climate risks, opportunities, and actions at national, sectoral, and local levels. This final chapter aims to provide guidance to policy makers in Tunisia on how best to move forward on this agenda. It has done this in two ways: (1) by providing a Framework for Action on Climate Change Adaptation, and (2) by putting forward a typology of policy approaches that are relevant to the region, to facilitate decision makers in formulating effective policy responses. Finally a policy matrix (see table 5.3) outlines key policy recommendations covered in each of the chapters for ease of reference.

Most of the actions aimed at increasing climate resilience recommended within this chapter will also have broader local development benefits by, for example, contributing to improved environmental governance, facilitating social inclusion, and sustainable growth. So, even in the absence of extreme events and high climate variability they are likely to present double-wins for Tunisian leaders.

Table 5.3 Policy Matrix

	Collect information on climate change adaptation and make it available	Provide human and technical resources and services to support adaptation	Provide assistance such as social protection for the poor and most vulnerable	Ensure a supportive policy and institutional framework	Build capacity to generate and manage finance and analyze financial needs and opportunities
General	– Establish a central bureau holding and disseminating data sets to concerned sectors on observed climate, socioeconomic characteristics, land use, and climate change scenarios.	– Identify needs (human and material). – Reinforce capacity for regional studies at the central level. – Put in place a process for regional consultations implicating nongovernmental organizations (NGOs). – Mainstream climate change adaptation into regional planning. – Put in place a system of monitoring and evaluation.	– Establish programs to support the basic needs and employment of the poor and most vulnerable.	– Support the implementation of the national strategy for adaptation. – Mainstream climate change considerations into the new constitution.	– Ensure mainstreaming of climate change at all levels and in all stages of planning and budgeting. – Put in place an institution responsible for coordinating access to external finance for climate change adaptation.
Climatology	– Climate scenario and impact analyses with other countries in the region (recognizing that many climate risks transcend state boundaries). – A MENA workshop to share lessons learnt and findings from climate studies across the Arab region as a whole.	– Enhance national and/ or regional capacity to utilize existing international programs on satellite retrievals and data. – Train and enhance capacity to work with and use comprehensive data sets such as reanalysis products. – Build capacity to use regional climate data information[a]	– Empower civil authority to be in charge of making data available for public use.	– Enhance regional collaboration on early warning systems, including use and dissemination of existing extended forecasts (available through WMO, etc.).	– Include climatological data collection in the national budget including costs related to data rescue; extending the number of weather stations.

(table continues on next page)

Table 5.3 Policy Matrix *(continued)*

	Collect information on climate change adaptation and make it available	Provide human and technical resources and services to support adaptation	Provide assistance such as social protection for the poor and most vulnerable	Ensure a supportive policy and institutional framework	Build capacity to generate and manage finance and analyze financial needs and opportunities
	– Collaborative research projects involving national and international experts addressing specific knowledge gaps: (1) coproduction and validation of climate scenarios for Tunisia; (2) enhanced capability for seasonal forecasting of drought at national levels and for agro-economic regions [for example, Normalized Difference Vegetation Index (NDVI), Standardized Precipitation Index (SPI), food price indices]; and (3) building technical capacity for impacts modeling. – Extend the coverage of the observational network in order to ensure a minimal station density to reflect climate variability and likely change in the country/region (also beneficial to weather forecasting and early warning systems).	– Promote and use available products for impacts and climate change risks analyses among users. – Establish regional/international centers of excellence. – Improve use of knowledge of centers existing within the country through staff exchange and through enhanced regional and international cooperation.	– Combine climate data with socio-economic data in order to obtain information that can assist in building resilience, including for the poor and most vulnerable.	– World Meteorological Organization (WMO) is promoting a new large-scale initiative on climate services. This initiative is crucially dependent on the active participation of the member states; capacities to explore and contribute to these efforts are critically needed. – Encourage international collaboration—wealth of satellite information can complement ground information	– Establish centers of excellence. – Build capacity and training.
Rural	– Assess changes in agricultural production levels/yields for indicator crops. – Model the food supply chains, model how they operate, and how they will be impacted by climate change.	– Develop knowledge and skills related to climate resilient agricultural practices such as growing salt-tolerant, heat-tolerant, and pest-resistant crop and livestock species, conservation agriculture, increasing irrigation efficiency and using nonconventional water resources.	– Use targeted transfers during price spikes and crop failures to support the most vulnerable. – Support development of access to markets for agricultural and other rural produce.	– Create a clear but coordinated governance structure to implement climate change adaptation measures at central and local levels across ministries responsible for agriculture, water, and the economy.	– Develop capacity to estimate the financial risks for not applying climate change adaptation and how to maximize risk management through available financial instruments.

(table continues on next page)

Collect information on climate change adaptation and make it available	Provide human and technical resources and services to support adaptation	Provide assistance such as social protection for the poor and most vulnerable	Ensure a supportive policy and institutional framework	Build capacity to generate and manage finance and analyze financial needs and opportunities
– Monitor state of water (groundwater and salinity levels) and soil conditions (depth and carbon content), and agricultural activities in "most at risk" agricultural zones (using indicator areas of marginal lands, rainfed areas from the four regions).	– Develop human and technical resources to optimize food chain systems particularly in transport, marketing, improving value-added developments, and establishing cooperatives.	– Support development of schools and training facilities to nurture both basic academic and vocational skills and provide necessary incentives to ensure attendance is possible.	– Develop a coordinated national policy, likely to be across ministries, supporting food security and rural livelihood developments, balancing risks with possibilities and mindful of water and energy security vulnerabilities. – Create farmers' associations that link directly to the Ministry of Agriculture and agricultural research/extension services to ensure clear flow of knowledge to all areas from top-down and bottom-up.	– Enhance capacity to assess all possibilities for meeting food demand balancing economics with geopolitical risks.

Source: World Bank data.

Note: MENA = Middle East and North Africa; NGOs = nongovernmental organizations; WMO = World Meteorological Organization.

a. For example by means of Geographical Information Systems (GIS).

Notes

1. For an overview of available tools to assist in climate risk analysis see http://climat-echange.worldbank.org/climatechange/content/note-3-using-climate-risk-screening-tools-assess-climate-risks-development-projects.

2. *Programme des écoles durable*

3. In 2010, Tunisia ranked 32nd out of 133 countries overall in the World Economic Forum Global Competitiveness Report. This was the top rank of any African country.

4. "MENA Assessment of the Local Manufacturing Potential for CSP Projects," Ernst & Young/Fraunhofer for World Bank/ESMAP, January 2011.

References

Barrett, C. B., B. J. Barnett, M. R. Carter, S. Chantarat, J. W. Hansen, A. G. Mude, D. E. Osgood, J. R. Skees, C. G. Turvey, and M. N. Ward. 2007. "Poverty Traps and Climate Risk: Limitations and Opportunities of Index-Based Risk Financing." IRI Technical Report 07–03, Working Paper, International Research Institute for Climate and Society, Columbia University, New York.

Ben Mansour, M. 2011. "Climate Prediction and Monitoring for Tunisia." National Oceanic and Atmospheric Administration (United States), National Center of Environnement Prediction, College Park, Maryland. http://library.wmo.int/pmb_ged/thesis/Tunisia-Ben_Mansour.pdf.

Burke, M., and D. Lobell. 2010. "Food Security and Adaptation to Climate Change: What Do We Know?" In *Climate Change and Food Security*, edited by D. Lobell and M. Burke, 133–53. Dordrecht, The Netherlands: Springer Science + Business Media, B.V.

Fattouh, B., and L. El-Katiri. 2012. "Energy Subsidies in the Arab World." Arab Human Development Report Research Paper Series, UNDP, Regional Bureau for Arab States, Oxford.

Hainoun, A. 2008. "Vulnerability Assessment and Possible Adaptation Measures of Agricultural Sector." United Nations Development Programme, Unpublished Report.

Martins-Filho, C., A. S. Taffesse, S. Dercon, and R. V. Hill. 2010. "Insuring Against the Weather: Integrating Generic Weather Index Products with Group-based Savings and Loans." Seed Project Selected, United States Agency for International Development, Washington, DC. http://i4.ucdavis.edu/projects/seed%20grants/MartinsFilho-Bangladesh/files/Martins-Filho%20proposal%20short.pdf.

OECD (Organisation for Economic Co-operation and Development). 2009. "Policy Guidance on Integrating Climate Change Adaptation into Development Cooperation." Organisation for Economic Cooperation and Development (OECD), Paris. http://www.oecd.org/document/40/0,3343,en_2649_34421_42580264_1_1_1_1,00.html.

Qiblawey, H., F. Banat, and Q. Al-Nasser. 2011. "Performance of Reverse Osmosis Pilot Plant Powered by PV in Jordan." *Renewable Energy* 36 (12): 3452–60.

Verner, D., ed. 2012. "Adaptation to a Changing Climate in the Arab Countries: A Case for Adaptation Governance and Leadership in Building Climate Resilience." MENA Development Report, World Bank, Washington, DC.

WMO (World Meteorological Organization). 2002. "Report of the Climate Database Management Systems Evaluation Workshop." WCDMP Report 50, Annex 10, Geneva, Switzerland. http://www.wmo.int/pages/prog/wcp/wcdmp/wcdmp_series/documents/Annex10.pdf.

World Bank. 2009. "Doing Business 2010. Reforming Through Difficult Times." World Bank, Washington, DC. http://www.doingbusiness.org/reports/global-reports/doing-business-2010/#sub-menu-item-link.

List of Daily Weather Stations in NCDC Archive that Were Reporting in 2011

WMO reference	Station name	Latitude (°N)	Longitude (°E)	Elevation (m)	Years
607230	Beja	36°44′	9°11′	158	1988
607140	Bizerte	37°15′	9°48′	5	1943
607690	Djerba Mellita	33°52′	10°46′	4	1966
607800	El Borma	31°41′	9°11′	258	1983
607650	Gabes	33°53′	10°06′	5	1957
607450	Gafsa	34°25′	8°49′	314	1957
607403	Habib Bourguiba Int	35°46′	10°45′	2	1989
607250	Jendouba	36°29′	8°48′	143	1943
607350	Kairouan	35°40′	10°06′	68	1957
607390	Kasserine	35°10′	8°50′	707	2001
607640	Kebili	33°42′	8°58′	46	1988
607200	Kelibia	36°51′	11°05′	29	1943
607320	Lekef	36°08′	8°42′	518	1989
607420	Mahdia	35°30′	11°04′	12	2003
607700	Medenine	33°21′	10°29′	116	1977
607400	Monastir-Skanes	35°40′	10°45′	2	1943
607750	Remada	32°19′	10°24′	300	1957
607500	Sfax El-Maou	34°43′	10°41′	21	1960
607480	Sidi Bouzid	35°00′	9°29′	354	1978
607340	Siliana	36°04′	9°22′	443	1980
607100	Tabarka	36°57′	8°54′	20	1974
607380	Thala	35°33′	8°41′	1091	1977
607600	Tozeur	33°55′	8°10′	86	1973
607150	Tunis-Carthage	35°50′	10°14′	3	1943
607255	Zine El Abidine Ben	36°05′	10°26′	21	2010

Source: World Bank data.

Notable Floods in Tunisia (Dartmouth Flood Observatory)

Location	Date	Duration (days)	Fatalities	Displaced	Damage (US$)	Notes
Sabalet Ben Ammar area. Bizerte. Tunis area. Ariana, Manouba	13-Oct-07	4	13	0	0	Heavy rains cause flash flooding in wadis (dry river beds). 8 dead and 8 missing in Sabalet Ben Ammar.
Jandouba, Benzert	12-Dec-03	2	7	0		Heaviest rains in 30 years on December 12, more than 7 inches along the coast of of Tunisia. Flooding in valleys.
Tunis area	16-Sep-03	3	4	0		Torrential rains for days. Many areas of Tunis submerged. Heaviest rains in 50 years. On September 24 torrential rains returned to Tunis, causing more flash floods.
Northern Tunisia— Siliana region	25-Jan-03	2	2	0		Heavy rains have continued in Tunisia since DFO-015 in mid January.
Northern and central Tunisia—Sbeitl a province, Le Kef province. Jendouba, Béja, Manouba and Bizerte. Kasserine, Kairouan, Sidi Bouzid, le Kef and Monastir.	14-Jan-03	3	8	27,000		Heavy rains for a week in Tunisia following four year drought. Rivers bursting banks in the northwest. Worst floods in 10 years.

(table continues on next page)

Location	Date	Duration (days)	Fatalities	Displaced	Damage (US$)	Notes
Provinces: Kairouan, Sfax, Kasserine, Gafsa, Sidi Bou Zid, Tozeur, Nefta, Gabes	20-Jan-90	13	25	152,000	233 million	
Cap Bon region of northern Tunisia: Tunis, Governorates of Tunis, Ben Arous, Ariana, Zaghouan, Nabeul, Kairouan, Kasserine, Bizerte, Beja, Jeouba	30-Sep-86	4	20	500		

source: http://floodobservatory.colorado.edu/index.html.

Supplementary Tables and Figures for Chapter 3

Table C.1 Mathematical Presentation of the Dynamic Computable General Equilibrium Model—Core Model Equations

Production function	$Q_{ct} = a_{ct} \cdot \prod_f F_{fct}^{\delta_{fc}}$		(1)
Factor payments	$W_{ft} \cdot \sum_c F_{fct} = \sum_c \delta_{fc} \cdot P_{ct} \cdot Q_{ct}$		(2)
Import supply	$P_{ct} \leq E_t \cdot W_c^m \perp M_{ct} \geq 0$		(3)
Export demand	$P_{ct} \geq E_t \cdot W_c^e \perp X_{ct} \geq 0$		(4)
Household income	$Y_{ht} = \sum_{fc} \theta_{hf} \cdot W_{ft} \cdot F_{fct} + r_h \cdot E_t$		(5)
Consumption demand	$P_{ct} \cdot D_{hct} = \beta_{hc} \cdot (1 - v_h) \cdot Y_{ht}$		(6)
Investment demand	$P_{ct} \cdot I_{ct} = \rho_c \cdot \left(\sum_h v_h \cdot Y_{ht} + E_t b \right)$		(7)
Current account balance	$w_c^m \cdot M_{ct} = w_c^e \cdot X_{ct} + \sum_h r_h + b$		(8)
Product market equilibrium	$Q_{ct} + M_{ct} = \sum_h D_{hct} + I_{ct} + X_{ct}$		(9)
Factor market equilibrium	$\sum_c F_{fct} = s_{ft}$		(10)
Land and labor expansion	$s_{ft} = s_{t-1} \cdot (1 + \varphi_f)$	f is land and labor	(11)
Capital accumulation	$s_{ft} = s_{t-1} \cdot (1 - \eta) + \sum_c \dfrac{P_{ct-1} \cdot I_{ct-1}}{k}$	f is capital	(12)
Technical change	$\alpha_{ct} = \alpha_{ct-1} \cdot (1 + y_c)$		(13)
Notes:			

(table continues on next page)

**Table C.1 Mathematical Presentation of the Dynamic Computable General Equilibrium Model—
Core Model Equations** *(continued)*

	Subscripts		*Exogenous variables*
c	Commodities or economic sectors	*b*	Foreign savings balance (foreign currency units)
f	Factor groups (land, labor, and capital)	*r*	Foreign remittances
h	Household groups	*s*	Total factor supply
t	Time periods	*w*	World import and export prices
	Endogenous variables		*Exogenous parameters*
D	Household consumption demand quantity	α	Production shift parameter (factor productivity)
E	Exchange (local/foreign currency units)	β	Household average budget share
F	Factor demand quantity	γ	Hicks neutral rate of technical change
I	Investment demand quantity	δ	Factor input share parameter
M	Import supply quantity	η	Capital depreciation rate
P	Commodity price	θ	Household share of factor income
Q	Output quantity	κ	Base price per unit of capital stock
W	Average factor return	ρ	Investment commodity expenditure share
X	Export demand quantity	υ	Household marginal propensity to save
Y	Total household income	φ	Land and labor supply growth rate

Source: World Bank data.

Table C.2 Social Accounting Matrix (SAM) Disaggregation

Activities & Commodities	*Factors*	*Institutions*
Wheat	Family workers	Enterprises
Other cereals	Agricultural workers	Rural farm households
Legumes	Non-agricultural workers	Rural non-farm households
Forage crops	Capital	Urban households
Olives	Rain-fed land	
Other fruits	Irrigated land	**Other**
Vegetables	Perennial land	Government
Other agriculture		Direct taxes
Livestock		Import tariffs
Forestry		Savings-Investment
Fishing		Rest of world
Meat		
Milk and its products		
Flour milling & its products		
Oils		
Canned food products		
Sugar and its products		
Other food products		
Beverages		
Other manufacturing and non-manufacturing industries		
Services		

Source: World Bank data.

Figure C.1 Global Effects of Climate Change on Households

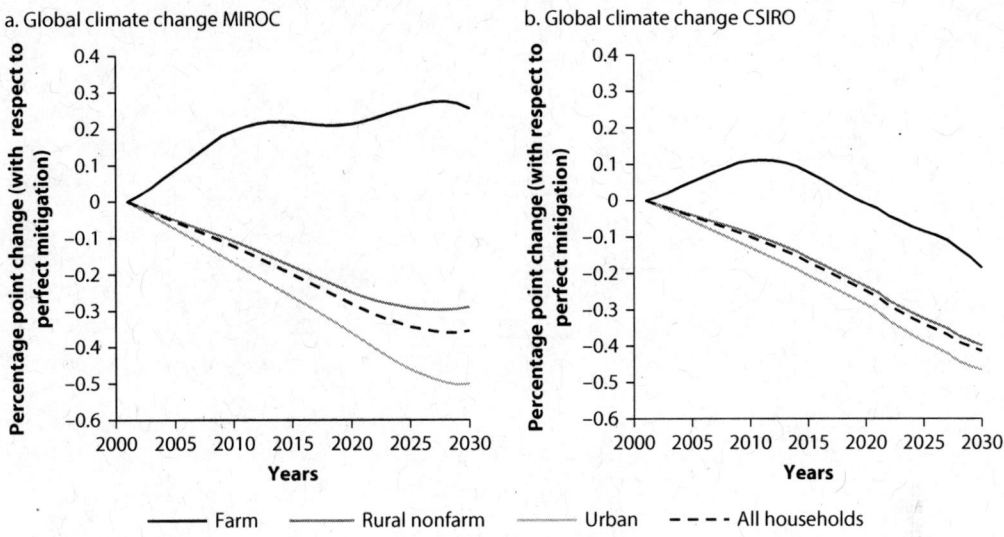

a. Global climate change MIROC

b. Global climate change CSIRO

Farm ——— Rural nonfarm ——— Urban - - - - All households

Source: World Bank data.

Figure C.2 Local Climate Change Effects on Households

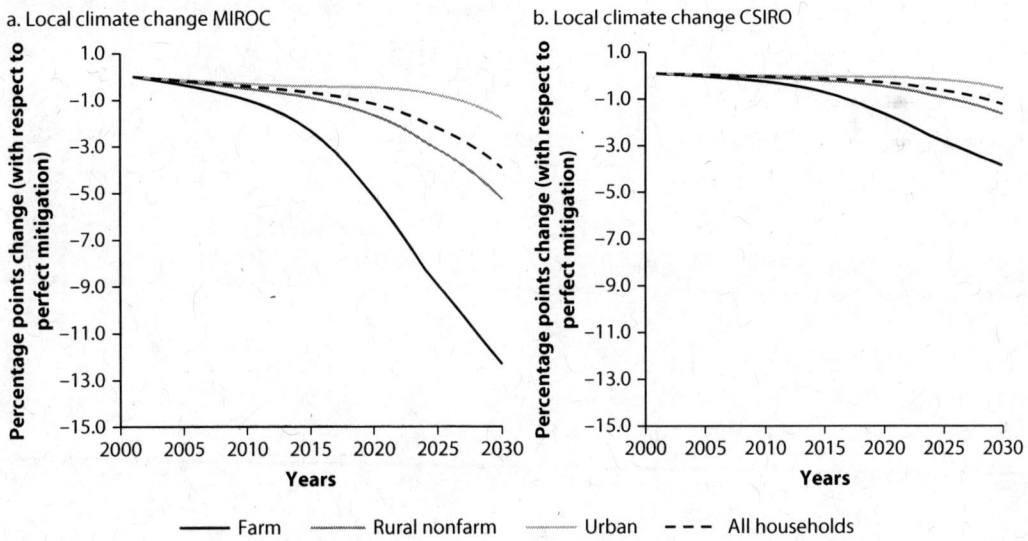

a. Local climate change MIROC

b. Local climate change CSIRO

Farm ——— Rural nonfarm ——— Urban - - - All households

Source: World Bank data.

Analytical Framework and Field Strategy

Social Dimensions and Analytical Framework

The objective of the fieldwork is to further understand the social dimensions of climate change among the rural population of Tunisia. Do people understand this the same way here? How do they perceive the effects of climate change in rural Tunisia; how and with what do they respond? At a global and regional scale the effects of climate change are regionally and socially unequal. While people in the upper end of the per-capita income distribution continue to enjoy the benefits of high-carbon economies, millions of people in middle- and low-income countries have their well-being, homes, and livelihoods threatened by climate change, despite their negligible contribution to the causes of climate change. In the Maghreb region, the negative impacts of climate variability and change are already impacting people's lives and livelihoods, as well as hitting poor people and regions the hardest. The poor are often the most vulnerable because they are heavily dependent on natural resources, they tend to live in marginal areas that are prone to drought and flood, they are often migrants, they are less educated, and some are ethnic minorities with less access to social safety nets. Because they have fewer assets to use as a buffer, it is also much more difficult for the poor to adapt their livelihoods to climate variability and to cope with extreme climate events. Asset-poor rural communities in the arid areas of the Maghreb, for instance, have generally few resources and little capacity to adapt to the changing climate.

People and governments are increasingly at a loss as to what to expect regarding the climate, and hence what decisions to make, for example, regarding agricultural activities, particularly given changes in the timing and intensity of rainfall. A key asset of poor farmers is their traditional knowledge about their environment, and that knowledge may no longer be reliable without the support of forecasting technology and additional climate information.

Climate change is a threat to short- and long-term development because it restricts human potential and freedom while reducing the capabilities that empower people to make choices about their well-being and livelihood (Mearns, Norton, and Cameron 2010, Verner 2010). As the Stern Review (2007) argues, it is paramount for climate change to become fully integrated into development policy, and for international support to be increased. Economic, human, and social development is key to the effort to reduce potential conflicts, migration, and loss of livelihood systems, and to improve infrastructure to enable people and communities to cope with climate change.

Climate variability and change are superimposed on existing risks and vulnerabilities that poor and marginalized groups typically face. The social impacts of climate change depend not only on biophysical factors, but on the vulnerability of people and institutions. Vulnerability to climate variability and change includes susceptibility to harm from exposure and a lack of adaptive capacity to cope with environmental change (Adger 2006). Climate variability and change erode livelihood assets by compounding existing vulnerabilities. These vulnerabilities are determined by socioeconomic factors such as an individual's age, gender, and ethnicity; a household's asset base and degree of integration with the market economy; as well as a community's capacity to tap into social capital among local residents and draw on national support systems to enhance resilience to climate change.

The livelihoods and well-being of people living in The Maghreb are challenged by a number of interrelated stress factors that are caused, or may be enhanced by, climate change:

- Stress on rural income generation and employment
- Stress on food security, both at the household and national level.
- Stress on natural resources, particularly water
- Stress on people's health.

Excessive stresses can lead to migration if people lack the assets to cope with them.

To further understand these stresses and to permit a systematic analysis of the effects of climate change on the poor and vulnerable, we combine several qualitative and quantitative fieldwork techniques informing an analytical framework successfully used in Latin America (2008–10) (Kronik and Verner 2010) and Syria (2010–11) (Verner, forthcoming). The framework is a slightly adapted version of the Department for International Development's Sustainable Livelihoods Framework (SLF) (Department for International Development (UK) 2001). This framework enables us to see how different aspects of climate change and climatic variability affect people's assets, their livelihood strategies, and livelihood outcomes, and hence their wellbeing. It also helps to identify entry points for how to raise income, increase well-being, reduce vulnerability, improve food security, and achieve more sustainable use of resources.

To understand people's vulnerability and to assess their livelihood outcomes, the SLF focuses on five factors:

- **Vulnerability context** refers to the environment in which people live. All types of external trends and shocks, including seasonality and climatic variability, affect people's livelihoods and the wider availability of assets. Gradual climate change and global warming may produce a wide range of shocks, such as flooding, hurricanes, tornadoes, and droughts, which in turn may cause negative outcomes such as changes in economic status and health that may lead to conflict, migration, or other problems. Climate change may also affect people's vulnerability by altering seasonality (for example of rainfall) and increasing climatic variability.
- **Livelihood assets** of individuals, households, or communities include physical, financial, human, social, natural, and cultural capital (table D.1). The more assets available to a person, the less vulnerable he or she is. Access to livelihood assets determines a person's level of resilience and adaptive capacity with respect to climate change. The reverse relationship is just as important; climate variability and change can affect access to assets. For this study, we have added cultural capital (Bourdieu 1973, 1986, 1996) to the five livelihood assets considered in DFID's original SLF. Our field research revealed that, particularly for indigenous people, the cultural dimension of livelihood strategies and social institutions is important for understanding the impacts of climate change and climatic variability.
- **Transformational structures and processes** at play within the community are the institutions, organizations, policies, and legislation that shape livelihoods. They operate at all levels, from the household, community, and municipality to the national and international, and in all spheres, from the most private to the

Table D.1 Definition of Livelihood Assets

Capital	Assets
Physical	The stock of plants, equipment, infrastructure, and other productive resources owned by individuals, the business sector, or the country itself.
Financial	The financial resources available to people (savings, supplies of credit).
Human	Investments in education, health, and the nutrition of individuals. Labor is linked to investments in human capital; health status determines people's capacity to work, and skill and education determine the returns from their labor.
Social	An intangible asset, defined as the rules, norms, obligations, reciprocity, and trust embedded in social relations, social structures, and societies' institutional arrangement. It is embedded at the micro-institutional level (communities and households) as well as in the rules and regulations governing formalized institutions in the marketplace, political system, and civil society.
Cultural	The knowledge, experience, and/or connections people have had throughout their lives.
Natural	The stock of environmentally provided assets such as soil, atmosphere, forests, minerals, water, and wetlands. In rural communities land is a critical productive asset for the poor, while in urban areas, land for shelter is also a critical productive asset.

Source: Adapted from DFID 2001.

most public. They effectively determine access to various types of assets, livelihood strategies, decision-making bodies, and sources of influence; the terms of exchange between different types of assets; and the returns (economic and otherwise) on any given livelihood strategy. They also directly affect people's sense of inclusion and well-being. These structures and processes can amplify vulnerabilities or be harnessed to enhance adaptive capacity and resilience.

- *Livelihood strategies* are influenced by vulnerability, assets, structures, and processes. The availability of choices in livelihood strategies is important, because they provide people with opportunities for self-determination and the flexibility to adapt over time. Adaptability is most likely to be achieved by improving poor people's access to assets—the building blocks for livelihood strategies—and by making the structures and processes that transform these into livelihood outcomes more responsive to their needs.
- *Livelihood outcomes* are the result of different livelihood strategies given specific vulnerabilities, assets, structures, and processes. Livelihood outcomes feed back into available assets, creating either a virtuous or vicious circle.

Field Strategy

Step 1: Country and Province Capitals:

- Meet key institutions and staff incl. the DGEQV; introduce objectives and briefly focus on methodology and timing.
- Ask into the existence and relative importance of transformational structures and processes at play within the community, through
- Venn institutional analysis: (1) national level, and then (2) provincial/governorate level in for example the southern governorates for agro-pastoralist livelihoods, as well as to the center-west governorates regarding agricultural livelihoods,
- Ask into important policies and legislation and the likelihood of important changes during the Constitutional Assembly and coming governments.

Ask into the state of main the physical, financial and natural capitals.

- The stock of plant, equipment, infrastructure and other productive resources
- The financial resources available to people (savings, supplies of credit, and so on)
- The stock of environmentally provided assets such as soil, atmosphere, forests, minerals, water, and wetlands.

Ask for relevant contact info, for further information and logistical arrangements.

Step 2: Approx. 20 Community visits in Southern governorates and if time allows in the Center-West governorates (SEE INTERVIEW FORMATS APPENDIX E)

Step 2: Approx. 20 Community visits in Southern governorates and if time allows in the Center-West governorates (SEE INTERVIEW FORMATS APPENDIX E) Sub-step 2.1: 1–2 interviews with community leader – entry point to introduce objectives and briefly focus on methodology and timing:

(A) **Local context,** socially differentiated relations, networks, and level and kinds of vulnerability and livelihood in 2010, 1990, 2030 through shocks and levels of vulnerability

(B) Ask into the existence and relative importance of **transformational structures** and processes at play within the community, through

- Venn institutional analysis
- Ask into important policies and legislation.

C1-6 Ask into the state of the main physical, financial, and natural capitals

- The stock of plant, equipment, infrastructure, and other productive resources
- The financial resources available to people (savings, supplies of credit, and so on)
- The stock of environmentally provided assets such as soil, atmosphere, forests, minerals, water, and wetlands.

(D1) Ask the local authority to kindly establish a **focus group meeting** with men and women with different experiences.

(D2) Obtain list of household names (heads of household—and copy to visual planning cards

(D3) Achieve rapport and points of entry for further information (phone, e-mail, and so on)

(D4) Ask for interview-persons, who have different perspectives (snowballing) for 1-hour interviews.

Sub-step 2.2: Individual interviews with 3–4 community members selected through maximum variation technique: (SEE INTERVIEW FORMATS – ANNEX A)

- (a) Enrich community level analysis through questions into the local context, socially differentiated relations, networks, and level and kinds of vulnerability and livelihood in 2010, 1990, 2030 through vulnerability/well-being analysis
- (b) Shocks (drought): types, intensity, frequency: Interviews open-ended and closed Q1: 2010, 1990, 2030 and Q2: 2010, 1990, 2030
- (c) Human capital: Investments in education, health, and the nutrition of individuals. Open-ended Q1: 2010, 1990, 2030 and closed (high/medium/low) Q2: 2010, 1990, 2030

- (d) Social capital: rules, norms, obligations, reciprocity, and trust embedded in social relations, social structures, and societies' institutional arrangement. Open-ended Q1: 2010, 1990, 2030 and closed (high/medium/low) Q2: 2010, 1990, 2030
- (e) Cultural capital: The knowledge, experience and connections people have had throughout their lives, which enable them to succeed better than someone without such a background. Open-ended Q1: 2010, 1990, 2030 and closed (high/medium/low) Q2: 2010, 1990, 2030.

Sub-step 2.3: Focus group interviews – where possible.

Scenario analysis with focus groups:

1. The first step in this exercise is to develop a realistic climate change impact scenario for each province, which can

   (a) be related to a specific climate event and/or climate change tendency

   (b) be recalled and related to by most members of the community.

   The scenario will be related to a high level of specificity with regards to relevant aspects/themes, however interviews will be held in an open-ended fashion, which will not provoke biases or responses designed to please the interviewer.

2. The second step entails the establishment of carefully selected focus groups with representative segments from the community based on factors such as gender, social group, and livelihood strategy.
   The scenario will be laid out, and discussions will be facilitated and recorded in answer to such questions as:

   "What did people do when they experienced the climate change event?"

   "Which resources (assets) did they draw upon?"

   "Which assets were temporarily/finitely damaged?"

   "Which resources (assets) could they substitute with?"

   "What do you think would have improved your ability to cope and adapt during that climate event?"

   The consultant will inquire further about the institutions/networks that some individuals/families were able to draw upon to increase the effect of their coping strategies and relate these answers to the stakeholder/institutional mapping.

Step 2.4: Interviews with Rural Communities
Open-ended interviews covering context, shocks, capitals, and structures.

Guidelines for field researchers:
Four different information-gathering techniques will be applied in rual communities to establish the range and types of information needed to support the various areas of the report:[1]

Well-being ranking to reach a first social mapping and local perceptions of importance of assets and opportunity structures: Key informants (3–4 per community) will individually rank households according to well-being. The key informant will be asked what is common to the households in each chosen rank, and what the differences between the ranked groups are. The explanations will be noted down (and later indicators will be drawn out) as well as the numbers and the rank (typically from three ranked groups). The indicators and assets will inform the necessary modification of the interview guide, to ensure it relates to local context.

Institutional/stakeholder mapping—Venn diagram: Key informants will first list all relevant institutions (informal, local, provincial, national, public, and private). A card will be written for each of the institutions. The cards will be placed with relative distance to the center—the center reflecting most valuable/trustworthy when ability to mobilize assets is important. Issues of differentiated levels and types of vulnerability will be explored and brought into the third exercise.

Scenario analysis with focus groups: The first step in this exercise is to develop a realistic climate change impact scenario for each province which can (a) be related to a specific climate event and/or climate change tendency, and (b) be recalled and related to by most members of the community. The scenario will be related to a high level of specificity with regards to relevant aspects/themes, however, interviews will be held in an open-ended fashion, which will not provoke biases or responses designed to please the interviewer. The second step entails the establishment of carefully selected focus groups with representative segments from the community based on factors such as gender, social group, and livelihood strategy. The scenario will be laid out, and discussions will be facilitated and recorded in answer to such questions as "What did people do when they experienced the climate change event?"; "Which resources (assets) did they draw upon?"; "Which assets were temporarily/finitely damaged?"; "Which resources (assets) could they substitute with?"; and "What do you think would have improved your ability to cope and adapt during that climate event?" The consultant will inquire further about the institutions/networks that some individuals/families were able to draw upon to increase the effect of their coping strategies and relate these answers to the stakeholder/institutional mapping.

Key informant interviews to fill gaps of understanding. Several techniques can be employed to establish a maximum variation sample depending on what is deemed relevant and time efficient.

Draft interview Guide: Semi-structured Questionnaire

Semi-structured questionnaire to be used with key informants in the Tunisian case-study on Climate Change and rural communities and to be modified for case studies among rural communities in Tunisia and potentially a third country. Questions will be adapted according to the type of key informant (for instance, which level of administration he/she represents).

1. Where were you during recent drought (or other climate event)?
2. How did you and your family react?
3. Which means did you use to respond to/cope with the situation?
4. Do you know of people in the community or nearby communities that reacted (adapted) differently than you and your family (or your community)?
5. Why do you think they acted (adapted) differently?
6. Can you classify the people in your community (or the communities in your area) in at least three groups that reacted/adapted differently? What are the elements that distinguish these three groups?
7. How long do you think each group (or your community or the communities in your area) took (or will take) to return to the state they were in before the drought? Why?
8. Do you think that people from your community would react (or reacted) differently than the people in the urban areas? Why?
9. How would you react if a drought happened today? Do you think that the natural resource base, which is the source of community livelihood, would recover faster or slower than if the event had occurred 20 years ago? Why?
10. Which types of conflicts arise or are strengthened during or after droughts?
11. Has anybody migrated permanently caused by the droughts? Why, why not?

Questions for Scenario Exercise:

The scenario will be played out ("in 20 years there will be more continuous dry days; rain will be more intense; the storms may be stronger..."), and discussions facilitated and recorded, asking into:

1. What did people do when they experienced the CC event?
2. Which resources (assets) did they draw upon?
3. Which assets where temporarily/finitely damaged?
4. Which resources (assets) could they substitute with?
5. What do you think would have improved your ability to cope and adapt during that climate event?

Venn Institutional Analysis:

1. Draw a circle and describe that the center is the place of most importance and less importance is further away from the center.

2. Which institutions were present during the drought? Please describe them briefly.
3. Which was the most important institution in your opinion for supplying information, relief, and rehabilitation?
4. Please order the rest of the institutions according to importance for you and the community.

Table D.2 How Key Information Types Are Covered

What	Who	How	When/Where
Vulnerability context			
–Situation/livelihood Q1: 2010, 1990, 2030 Q2: 2010, 1990, 2030	Community members: Max-variation sample, snowballing until saturation	**Vulnerability and or well-being ranking** and/or open-ended interviews	3–4 interviews per community, men & women
–Shocks (drought): types, intensity, freq. Q1: 2010, 1990, 2030 Q2: 2010, 1990, 2030	Community members individually or in focus groups if possible	**Scenario analysis**	4–8 interviews per community, men & women Focus groups if possible
Livelihood assets of individuals, households, or communities—capitals:			
Physical capital Q1: 2010, 1990, 2030 Q2: 2010, 1990, 2030	Community leaders and members individually	Structured interview— closed questions	4–8 interviews per community Badia commission
financial capital: Q1: 2010, 1990, 2030 Q2: 2010, 1990, 2030	Community leaders and members individually	Structured interview— closed questions	1–2 interviews per community Province/state authority
Human capital: Q1: 2010, 1990, 2030 Q2: 2010, 1990, 2030	Community members individually	Open-ended and structured interview	4–8 interviews per community
Social capital: Q1: 2010, 1990, 2030 Q2: 2010, 1990, 2030	Community members individually	Open-ended and structured interview	4–8 interviews per community
Natural capital: Q1: 2010, 1990, 2030 Q2: 2010, 1990, 2030	Community leaders and members individually	Structured interview— closed questions	4–8 interviews per community Province/state authority
Cultural capital: livelihood strategies & social institutions Q1: 2010, 1990, 2030 Q2: 2010, 1990, 2030	Community members individually	Open-ended and structured interview	4–8 interviews per community
Transformational structures and processes at play within the community - institutions, organizations, policies, and legislation that shape livelihoods			
Institutions, organizations, access decision making	Community members: Max- variation sample, snowballing until saturation	Venn diagram & open-ended interviews	4–8 interviews per community, men & women
Policies, and legislation	– Community leaders – Province/state authority	Interviews	1–2 interviews per community Province/state authority

Source: World Bank data.
Note: Q1: Qualitative and Q2: Quantitative

Note

1. The aim is not to establish statistical impact data.

References

Adger, W. N. 2006. "Vulnerability, Global Environmental Change 16." Tyndall Centre for Climate Change Research, School of Environmental Sciences, University of East Anglia, Norwich. 268–81.

Bourdeau, P. 1973. "Cultural Reproduction and Social Reproduction." In *Knowledge, Education and Cultural Change*, edited by R. Brown.. London: Willmer Brothers Limited.

———. 1983, 1986. "The Forms of Capital." In *Handbook of Theory and Research for the Sociology of Education*, edited by J. G. Richardson, 241–58. New York: Greenwood Press.

Department for International Development (UK). 2001. "Sustainable Livelihoods Guidance Sheet." http://www.nssd.net/pdf/sectiont.pdf.

Kronik, J., and D. Verner. 2010. *Indigenous Peoples and Climate Change in Latin America and the Caribbean*. Washington, DC: World Bank.

Mearns, R., A. Norton, and E. Cameron. 2010. *The Social Dimensions of Climate Change. Equity and Vulnerability in a Warming World*. Washington DC: World Bank, p. 232.

Stern, N. 2007. *Stern Review on the Economics of Climate Change*. Cambridge, UK: Cambridge University Press.

Verner, D., ed. 2010. *Reducing Poverty, Protecting Livelihoods, and Building Assets in a Changing Climate: Social Implications of Climate Change Latin America and the Caribbean*. Washington, DC: World Bank.

———. Forthcoming. *Syria Rural Development in a Changing Climate*. Washington, DC: World Bank.

Questionnaire Format

Questions pour le Groupe

Date de l'Interview: _____/_____ 2011 Heure: _____h_____
Nom de la communauté: _____ Zone:_____
Gouvernorat: _____

A) Contexte local, relations socialement différenciées, réseaux, et niveaux et types de vulnérabilités et moyens de vie dans le passé, maintenant et à l'avenir.	
A1)	**Chocs (sécheresse):** Comment étiez-vous préparés pour une sécheresse prolongée dans le passé ? Comment êtes vous préparés pour une période de sécheresse prolongée à l'avenir, en tenant du tenant compte de la situation actuelle ?

A2)	Chocs (sécheresse):	Bien	Moy	Bas
passé	Comment étiez-vous préparés pour une sécheresse prolongée dans le passé ?			
maint	Comment êtes-vous préparés pour une sécheresse prolongée maintenant ?			
avenir	Comment êtes vous préparés pour une période de sécheresse prolongée à l'avenir ?			

C) Stratégie pour les moyens de vie et les actifs clés:

C1a) PHYSIQUE

Comment se présente l'accès **aux moyens de transport** dans le passé, maintenant et à l'avenir ?

Dans le passé:

Maintenant:

A l'avenir:

C1a)	PHYSIQUE : L'accès **aux moyens de transport** dans le passé, maintenant et à l'avenir ?	Bon	Moy	Mauv
passé				
maint				
avenir				
C3a)	ENVIRONNEMENTAL: Comment se présente l'accès à **l'eau**, dans le passé, maintenant et à l'avenir ?			
	Dans le passé:			
	Maintenant:			
	A l'avenir :			

C3a)	ENVIRONNEMENTAL : L'accès à **l'eau** (bon/moyen/mauvais) dans le passé, maintenant et à l'avenir ?	Bon	Moy	Bas
passé				
maint				
avenir				

C3b)	ENVIRONNEMENTAL: Comment se présente l'état du **pâturage**, dans le passé, maintenant et à l'avenir ?			
	Dans le passé: **Maintenant:** **A l'avenir :**			

C3b)	ENVIRONNEMENTAL: L'accès au pâturage (bon/moyen/mauvais) dans le passé, maintenant et à l'avenir ?	Bon	Moy	Mauv
passé				
maint				
avenir				

C2a)	FINANCIER: Comment se présente l'accès aux sources de revenus NON-RELIEES A l'ELEVAGE dans le passé, maintenant et à l'avenir ? **(OPTIONNEL)**			
	Dans le passé: **Maintenant :** **A l'avenir :**			

C2a)	FINANCIER : L'accès aux sources de revenus NON-RELIEES A l'ELEVAGE DE MOUTONS (bon/moyen/mauvais) dans le passé, maintenant, et à l'avenir ?	Bon	Moy	Mauv
passé				
maint				
avenir				
C5e)	SOCIAL : D'après vous – comment la période de sécheresse a-t-elle affectée le niveau de confiance entre les gens de la localité?			
C5e)	SOCIAL : D'après vous – comment la période de sécheresse a-t-elle affectée le niveau de confiance entre les gens de la localité?	Bon	Moy	Mauv
passé				
maint				
avenir				
C6)	**CULTUREL:** **Le savoir, l'expérience et les connections que les gens ont eu au cours de leurs vies, qui leur a permis de réussir mieux que quelqu'un sans un tel contexte.**			
C6a)	En vue de la période de sécheresse que vous vivez, quels sont vos acquis suite à votre mode de vie ? Et comment cela se compare-t-il à une famille de la ville ?			

		Elevé	Moy	Bas
C6e)	Comparez le niveau de connaissances et de savoir-faire en milieu rural pour résister à une période de sécheresse, dans le passé, maintenant, et à l'avenir.			
passé				
maint				
avenir				
C6d)	Est-ce que la pensée des âgés et des jeunes sont la même ici ?			
A3)	Vulnérabilité : Dans votre localité, pouvez décrire ce qui caractérise les ménages les plus vulnérables à la sécheresse, les ménages moyennement vulnérables, et les ménages qui ne sont pas vulnérables ?			
	1. Les plus vulnérables: **2. Vulnérabilité moyenne:** **3. Les moins vulnérables:**			

Questions Individuelles

Date de l'Interview: _____/_____ 2011 Heure: _____h_____
Nom de la communauté: _____ Zone:_____
Gouvernorat: _____
Nom de l'interviewer:_____ Nom du père :_____
Age: _____, Genre: Homme:_____, Femme_____
Niveau de scolarité: _____

C1a)	PHYSIQUE ; L'accès de la famille aux moyens de transport dans le passé, maintenant, et à l'avenir	Bon	Moy	Mauv
passé				
maint				
avenir				
C3a)	ENVIRONNEMENTAL : L'accès à **l'eau et au fourrage** (bon/moyen/ mauvais) dans le passé, maintenant et à l'avenir ?			
passé				
maint				
avenir				
C2a)	FINANCIER: Comment se présente l'accès de votre famille aux sources de revenus NON-RELIEES A L'ELEVAGE DE MOUTONS dans le passé, maintenant, à l'avenir ?			
Réponse	**Dans le passé:** **Maintenant:** **A l'avenir :**			

C2a)	FINANCIER	Bon	Moy	Mauv
passé	L'accès de votre famille aux sources de revenus NON-RELIEES A l'ELEVAGE DE MOUTONS (bon/moyen/mauvais) dans le passé, maintenant, et à l'avenir ?			
maint				
avenir				
C2b)	FINANCIER: Comment se présente l'accès de votre famille aux prêts dans le passé, maintenant, et à l'avenir ?			
	Dans le passé :			
	Maintenant :			
	A l'avenir :			

C2b)	FINANCIER	Bon	Moy	Mauv
passé	L'accès de votre famille aux prêts (bon/moyen/mauvais) maintenant, il y a 20 ans et dans 20 ans?			
maint				
avenir				

C4a)	CAPITAL HUMAIN: Décrivez l'évolution du niveau de scolarité au sein de votre famille?		Alph	Prim B	Sec	Sup	Dipl
		Homme					
		Femme					

C4b)	CAPITAL HUMAIN: Dans le passé quand vous étiez malade, est-ce que vous êtes allé à la clinique privée ou l'hôpital public ?			
Rép				
C4b)	**Etes-vous allé à un centre de santé ?**	Public	Privé	Aucun
	Dans le passé?			
	Actuellement ?			
	Et à l'avenir ?			
C5a)	**CAPITAL SOCIAL**	Facil	Pas Facil	Impossible
passé	Si vous avez besoin d'argent est-ce facile d'obtenir de l'aide– et cela a-t-il changé entre maintenant et auparavant ?			
maint				
Rép.	De qui?			
C6d)	Est-ce que votre façon de penser est partagée par un âgé/jeune ?			
C6e)	Comparez le niveau de connaissances et de savoir-faire en milieu rural pour résister à une période de sécheresse, dans le passé, maintenant, et à l'avenir.	**Elevé**	**Moy**	**Bas**
passé				
maint				

Groupde Vulnérabilité	Dans quel groupe de vulnérabilité est-ce que vous vous considérez ? **1. Les plus vulnérables:** **2. Vulnérabilité moyenne :** **3. Les moins vulnérables:**

ECO-AUDIT
Environmental Benefits Statement

The World Bank is committed to preserving endangered forests and natural resources. The Office of the Publisher has chosen to print World Bank Studies and Working Papers on recycled paper with 30 percent postconsumer fiber in accordance with the recommended standards for paper usage set by the Green Press Initiative, a non-profit program supporting publishers in using fiber that is not sourced from endangered forests. For more information, visit www.greenpressinitiative.org.

In 2010, the printing of this book on recycled paper saved the following:
- 11 trees*
- 3 million Btu of total energy
- 1,045 lb. of net greenhouse gases
- 5,035 gal. of waste water
- 306 lb. of solid waste

----

* 40 feet in height and 6–8 inches in diameter